D1005664

STEM the Tide

STEM the Tide

Reforming Science, Technology, Engineering, and Math Education in America

DAVID E. DREW

Foreword by Alexander W. Astin

The Johns Hopkins University Press
Baltimore

© 2011 The Johns Hopkins University Press
All rights reserved. Published 2011
Printed in the United States of America on acid-free paper
2 4 6 8 9 7 5 3

The Johns Hopkins University Press
2715 North Charles Street
Baltimore, Maryland 21218-4363
www.press.jhu.edu

Lyrics from "New Math" © 1965 Tom Lehrer, reprinted by permission of Tom Lehrer.

Library of Congress Cataloging-in-Publication Data

Drew, David E.
STEM the tide : reforming science, technology, engineering, and math education in America / David E. Drew. Foreword by Alexander W. Astin.
p. cm.
Includes bibliographical references and index.
ISBN-13: 978-1-4214-0094-5 (hardcover : alk. paper)
ISBN-10: 1-4214-0094-4 (hardcover : alk. paper)
1. Science—Study and teaching—Government policy—United States. 2. Technology—Study and teaching—Government policy—United States. 3. Engineering—Study and teaching—Government policy—United States. 4. Mathematics—Study and teaching—Government policy—United States. 5. Science and state—United States. 6. Technology and state—United States. I. Title.
Q183.3.A1D734 2011
507.1'073—dc22 2010048728

A catalog record for this book is available from the British Library.

Special discounts are available for bulk purchases of this book. For more information, please contact Special Sales at 410-516-6936 or specialsales@press.jhu.edu.

The Johns Hopkins University Press uses environmentally friendly book materials, including recycled text paper that is composed of at least 30 percent post-consumer waste, whenever possible.

For Cissy
and
for my Aunt Annette

Contents

Foreword

One of the distinguishing features of higher education in the United States is the frequency with which college students change their major fields of study. While undergraduates in most other countries of the world pretty much stick with their initial courses of study as they prepare for particular types of professions or careers, many American college students earn their degrees in a field other than the one in which they originally intended to major.

When we compare the initial major field choices of entering college freshmen in the United States with the fields in which they actually graduate, the big "losers" are what have come to be known as the "STEM" fields (science, technology, engineering, and mathematics). In other words, many fewer students end up earning degrees in STEM fields than initially intended to pursue a course of study in such fields when they started college. This "flight" from science and math actually begins well before college. Studies of middle and high school students have shown a steadily declining interest in math and science as students progress through the grades.

As far as sheer numbers of college students are concerned, the biggest "loser" among the various STEM fields is engineering, in part because engineering is such a popular initial career choice among new freshmen. In fact, if it weren't for all these engineering dropouts, net losses in fields like physics, math, and chemistry would be considerably larger, given that some of those who leave engineering end up majoring in other STEM fields. Most students who leave engineering, however, change to non-STEM fields like business, and since very few students switch *into* STEM fields from non-STEM fields during college, the large-scale picture is one of substantial overall losses for STEM—something approaching 50 percent. Net losses are even greater for women and for students from underrepresented racial and ethnic groups.

Opinions about the consequences of these trends are highly variable. Some observers believe that the substantial loss of students from the study of science and mathematics is not only inevitable but desirable: "These are very difficult and intellectually challenging fields, and it is important to weed out the less competent students to insure that only the smartest people are allowed to pursue careers in these fields." Others see these losses as potentially catastrophic for our country's businesses, industry, and educational system and as a threat to its global economic competitiveness. They also point out that the problems posed by this large-scale defection of students from the study of science are compounded by the fact that students in the United States have consistently performed below students in most other countries on tests of scientific and mathematical knowledge and competence.

I happen to believe that these two issues are intimately related, that the failure to acquire scientific knowledge and to master mathematical concepts in the early school years will cause many students to avoid the study of STEM fields during the college years. And even among those students who manage to retain their interest in STEM careers until they begin college, marginal preparation in science and math increases the likelihood that they will become frustrated and discouraged with college-level STEM courses and switch to other fields. At the same time, when so many students abandon the study of science and math, we deplete our resources of STEM talent needed not just for research and development, but also for educating future generations of students in these fields.

Why do these STEM losses happen? And why do our students perform so poorly in science and math when compared to students in other countries? Debates about these issues have typically pointed the finger at a dizzying array of "explanations": poor teaching, poor teacher training, teachers unions, poorly designed curricula, poorly run schools, low teacher pay, tightfisted politicians, inadequate testing, poorly motivated students, inadequate parental supervision, and so on. In this path-breaking book, David Drew looks at all of these theories with a critical eye, drawing together the best evidence from a wide variety of sources into a comprehensive picture of why we find ourselves in this position and what we can and should do about it.

Rather than relying on a single favorite explanation or solution, Drew convincingly outlines the full complexity of the problem and articulates a remedy that requires multiple strategies involving students, teachers, institutions, and policymakers. Readers may be surprised by the degree of attention given to beliefs and expectations, but Drew is able to marshal a considerable degree of evidence and argument in support of the idea that making significant improvements in science education in this country will first require that students, teachers and policymakers change their traditional beliefs about such things as "aptitude," and about who is and is not capable of mastering STEM subject matter. A growing body of evidence suggests that most students, given proper instruction and mentoring, would be able to achieve a significant level of competence in STEM fields. To accomplish such a goal will require the combined efforts of teachers; those who select and train teachers; schools, colleges and universities; and governmental agencies, especially the National Science Foundation.

This very readable book not only represents a comprehensive review of what is known from systematic research on STEM education, but is also replete with real life examples of specific reform efforts that have proven to be successful. It should be required reading for all science and math teachers, those who train the teachers, practicing scientists and engineers, and those who are responsible for funding research and setting science policy in our country.

Alexander W. Astin

Preface

As a first-year graduate student, I assisted in an evaluation of a mathematics reform program. In one way or another, I have been focused on STEM education ever since then. In this book I draw together much of what I have learned. More to the point, above and beyond my own modest contributions to this literature, I have scoured journals, books, reports, magazines, and newspapers to glean the key lessons about reforming and improving STEM education.

Over the years, as a manufacturing economy declined and was replaced by a high-tech global economy, this once remote intellectual outpost—how to most effectively teach math and science—has assumed critical importance.

I teach multivariate statistics to PhD students. Initially, I saw myself as a researcher who also taught. Gradually, I realized that my teaching and mentoring activities were as important a contribution, or more important. Furthermore, the deep learning about education that derives from committed teaching adds depth and understanding to otherwise abstract and disembodied analyses of charts, tables, and data about STEM education.

As I prepared this book, I benefited from suggestions made by a number of colleagues, including some of the creative educators whose work I highlighted. In addition to them, I benefited from conversations with Lourdes Arguelles, Ross Barrett, John Bear, Fitzgerald Bramwell, Jim Buckheit, Craig Cassidy, Samir Chatterjee, Karen Gallagher, Kenneth C. Green, Michael Howell, Maria Klawe, Bob Klitgaard, Jean Lipman-Blumen, Rob Lovelace, Pam Mason, Joe Maciariello, Tami Pearson, Jack Schuster, Daryl Smith, Gail Thompson, and Darryl Yong.

Several research assistants tracked down important articles and research studies for me. For this help I thank Franque Bains, Kathryn Flynn, Cherie Ichinose, Sara Kapadia, Laura Kazan, and Ming Zhou.

Two colleagues, Hedley Burrell and Paul Gray, read and critiqued an early draft of this manuscript. I also benefited from editorial suggestions made by Ashleigh McKown, my editor at the Johns Hopkins University Press; by anonymous reviewers who read the draft manuscript; and by Carolyn Moser, my copy editor. Ashleigh McKown proposed the title for this book.

I had outstanding secretarial assistance for this project from Nicole Jones, Kiriko Komura, and Teresa Wilborn.

I frequently collaborate with Martin Bonsangue, professor of mathematics at California State University at Fullerton. He and I coauthored the memorandum upon which the section of chapter 3 on evaluation is based. We also coauthored the technical report about STEM education in the Houston consortium upon which the section in chapter 6 is based. Dr. Bonsangue should rightfully be considered a coauthor of those two sections.

This volume began as a second edition of my book, *Aptitude Revisited: Rethinking Math and Science Education for America's Next Century*, also published by the Johns Hopkins University Press, in 1996. It rapidly became clear that there was enough fresh material to justify a new book. However, a few sections of that prior work are included in this book.

STEM the Tide

Introduction

> When I compare our high schools to what I see when I'm traveling abroad, I am terrified for our workforce of tomorrow.
>
> Bill Gates

Science, technology, engineering, and mathematics (STEM) education is vital preparation for today's high-tech information economy. Unfortunately, American students lag their counterparts in most other countries in achievement. This book presents a positive blueprint for reforming STEM education in our schools, colleges, and universities.

The United States (and, indeed, much of the world) is reeling from a severe economic crisis. Most adults define themselves in great part by their work, and suddenly, the form and demands of that work have changed. Simultaneously, Americans have become increasingly concerned about the dire state of education. American high school students consistently rank at or near the bottom in international assessments of educational achievement. With fewer qualified workers domestically, U.S. businesses are hiring more scientists, engineers, and other skilled workers from foreign countries.

The last 20 years have seen a dazzling array of innovations in technology, communications, financial products, and corporate structures, including the Internet, corporate outsourcing to remote locations throughout the world, and open source software platforms. But these innovations pose challenges to the status quo of an older, shrinking industrial economy.

Faced with new technological and educational challenges, many Americans have become discouraged about the future of the U.S. economy and about the diminishing career opportunities they, their children, and their grandchildren will face. Policy analysts and pundits have expressed anxiety, even fear, about whether the United States will continue to be a world leader economically. Scholars and think tanks have issued reports about globalization and have speculated about which countries and economies will dominate the next quarter century.

Each report has expressed concern about the state of American education and called for reform. But the next step—specifying exactly how that reform can be accomplished—has not yet been taken.

Focusing on Strengths, Not Weaknesses

In this book, I show that America has the knowledge and experience to reform our schools, colleges, and universities *now*. Our students can compete successfully in the global economy. American scientists and engineers can contribute creatively in this changed environment. For decades, my research has focused on the improvement of mathematics and science education; on building strong undergraduate programs; and on funding, managing, and strengthening university research programs. This book offers a primer on the latest research and discoveries of leading educators and social scientists about how to reform STEM education.

I am optimistic that we can fix American education and research now that we know what needs to be done. American students can be educated in elementary, middle, and high school, and in our colleges and universities, to compete successfully with their counterparts from other countries. American scientific leadership can be renewed by understanding how technology and commerce have changed in the past decade and by addressing the problems in American math and science education and in university research.

While we can learn from observing what works in other countries, it is a mistake for U.S. educators to slavishly imitate instructional strategies that work in other countries. We must build upon our unique strengths.

High Expectations, Limitless Potential

Many students have been excluded from the study of mathematics and science education because it has been falsely assumed that they lack the intelligence to master the subject. *This is the fatal flaw in American education.* By instituting consistently high expectations in these subjects, we can unleash a reservoir of hidden talent in the United States. Assumptions about predetermined intelligence and aptitude are wrong. All students—including women, students of color, and impoverished students—can master science and math and can be prepared now for America's future global economic challenges.

To keep high-tech jobs in the United States,

- we must keep talented students engaged in STEM education;
- we must evaluate rigorously any educational reform to ensure that it works;

- we must create an environment for talented teachers that fosters creativity and productivity;
- we must increase access to affordable, high-quality undergraduate education; and
- we must get students excited about scientific research.

STEM Education: The Historical Context

The first modern-era challenge to American technological leadership was the Soviet Union's successful launch of Sputnik in 1957, at the height of the Cold War. By today's standards, the satellite was tiny—just 22 inches in diameter—but it had a huge psychological and political impact on Americans. It circled the globe 560 miles up. The *New York Times* found space in the headline of its front-page story to note, "Sphere Tracked in 4 Crossings Over U.S."[1] It emitted an ominous and continuous beep, which could be heard by shortwave radio operators.

One response to the growing Soviet challenge was the National Defense Education Act (NDEA), which provided funding to improve mathematics and science education. (Incidentally, the word *defense* in the title of the act was superfluous, but in the midst of the Cold War that little word helped ensure passage of this bill.)

In the 1950s, the NDEA and the National Science Foundation funded a Yale University team called the School Mathematics Study Group to develop a new approach to teaching mathematics called "New Math." The curriculum introduced students to theoretical concepts, such as set theory, at a young age.[2] New Math yielded mixed results, and it did not succeed in reforming American mathematics education. Some felt that the emphasis on theory had the unintended consequence of reducing students' ability to carry out basic arithmetic problems, to say nothing of their parents' ability to assist them with their math homework. Harvard professor Tom Lehrer lampooned the new approach in his song "New Math":

> You can't take three from two,
> Two is less than three,
> So you look at the four in the tens place.
> Now that's really four tens,
> So you make it three tens,
> Regroup, and you change a ten to ten ones,
> And you add them to the two and get twelve,
> And you take away three, that's nine.
> Is that clear?

NDEA funding helped improve education, but much more support and improvement still was needed. Decades of math and science education reforms followed, some successful, some not.

In this book, I extract and present the best of what we have learned from experiments, demonstration projects, and studies about how to make our schools, colleges, and universities more effective. You will read about the work of creative Americans who are at the cutting edge of technology and education. Their innovative strategies and solutions can help repair American education and American research as well as build American technological and economic leadership in the twenty-first century. These forward-thinking individuals include:

- a superintendent who successfully led school systems in four of our nation's major cities;
- an astonishing high school teacher whose working-class students went on to transform the field of animation;
- a world-class inspirational mentor who has set the bar high for those who aspire to guide students; and
- a dynamic leader who, despite strong political opposition, founded a college that now prepares young people from all walks of life for technical careers.

America's Hidden Talent

> And I learned the greatest gift of all. The saddest thing in life is wasted talent. . . . And the choices that you make will shape your life forever.
>
> Calogero "C" Anello in *A Bronx Tale*[3]

The have-nots in American society—the poor, the disadvantaged, and people of color—are severely underrepresented in classrooms where mathematics and science are taught. Science education is vital for a technologically advanced society, but it is also a vehicle through which the inequalities of our society are perpetuated and exacerbated. If current trends continue, the proficiency gap in the sciences will widen between the haves and the have-nots, and this will damage our economy. In fact, the research reported in this volume strongly suggests that mathematics in particular is the crucial filter determining access to many prestigious, respected, and lucrative careers. Mathematics can be the catalyst for the social mobility of individuals and groups who have traditionally been outside the mainstream of the American economy. These individuals and groups represent a reservoir of hidden talent.

The factors that determine who receives mathematics and science education—and, thus, who has access to wealth and power in our society and our economy—are sociological and psychological. They are not matters of curriculum, innate ability, or school funding. These factors include teacher expectations, which affect both the options presented to students and how those students view themselves, or their "self-concept." Social scientists have carefully studied self-concept—what young people believe they are capable of achieving—and how it relates to actual learning, subject mastery, and goal-setting.

The data indicate that millions of people are erroneously discouraged from studying mathematics and science because of false assumptions about who has the ability to master these subjects. These assumptions become self-fulfilling expectations, which ultimately undermine the self-concept of female students, impoverished students, and students of color. Unfortunately, these erroneous assumptions are sometimes held by teachers or parents. But the resulting impacts on self-concepts, aspirations, and achievement are destructive and pernicious. If a person has been persuaded not to pursue his or her dream career as a result of being told, "You would need to study math, and let's face it, you're no good at that," then that person has suffered two injustices. The first is the establishment of an unjustified barrier to his or her professional aspirations. The second is the damage done to his or her self-concept.

Most people can master mathematics, even calculus, and other scientific fields. Every student must recognize this, and not let himself or herself be excluded from STEM education because of another person's erroneous judgments. Virtually every person is capable of receiving extensive, thorough, and rigorous training in STEM fields.

If, instead of excluding students, we expect and demand that virtually all students master mathematics and science, the strength of the American skilled labor force, and of the American economy, will increase.

The first step toward improving STEM education is to increase our nation's consciousness about the issues outlined above. Teachers, parents, and students themselves must recognize that virtually every child has the capacity to master mathematics and science; and all students should be taught these subjects. This is true for girls as well as for boys, for poor students and affluent ones, and for persons of every ethnicity. Beyond consciousness-raising, the research shows us how the reform of STEM education can be accomplished.

Fear of Math

There are many stereotypes about math:

- Asians are better at math than Americans.
- There's no need to learn math if you have a calculator and a computer.
- Most women can't do math.

These false statements have become stubborn myths in American culture. In a ranking of things that most Americans fear and hate, mathematics rates right up there with taxes and politicians.

Mathematician John Allen Paulos argues that Americans suffer from what he calls "innumeracy," the inability to deal with numbers and mathematics, a deficiency he describes as parallel to illiteracy. He observes that "unlike other failings which are hidden, mathematical illiteracy is often flaunted: 'I can't even balance my checkbook,' 'I'm a people person, not a numbers person' or 'I always hated math.' Part of the reason for this perverse pride in mathematical ignorance is that its consequences are not usually as obvious as are those of other weaknesses."[4] Even people who like mathematics often won't admit it. We've all met the die-hard sports fan who "hates math" but can tell you down to the third decimal place what Derek Jeter's slugging average will be if he gets extra base hits in his next two times at bat.

Avoidance of mathematics may explain the career decisions made by young people. Some who want to be doctors and dentists may choose other careers so that they will not have to take mathematics in college. I know an academic counselor at a large technical university who hears repeatedly, "I'd like to major in ——, but I can't do math, so I'm going to become a teacher instead." Think of the introduction the students of those future teachers will have to the joys of arithmetic and mathematics!

While many avoid mathematics, the research shows that those who master math tend to be most successful in college and beyond. Furthermore, virtually everyone can learn advanced mathematical concepts, even those who start late.

The "Strange Concept of Aptitude"

> As compared to students in high achieving countries, American students believe strongly that mathematical talent is innate, and believe less strongly that effort makes much difference.
>
> Anne C. Lewis, "Endless Ping-Pong over Math Education"[5]

A colleague of mine who is a mathematics professor in China has pointed out that virtually everybody in China learns advanced mathematics, while only select students in the United States do. In China, everyone is assumed to be able to master advanced concepts. Americans, he said, have a "strange concept of aptitude."

There are generally two approaches to teaching any subject matter: either the material is assumed to be understandable by most students and the challenge is devising an effective means of teaching that material, or the subject matter is believed to be so tough that only a few of the best and brightest will be able to learn it. Unfortunately, many mathematics teachers embrace the second philosophy. But mathematics is an accessible subject, understandable by most people in our society, not an arcane discipline only a few students can master.

Negative expectations held by teachers and counselors prevent many young people from taking the math courses they need. Teachers' attitudes and expectations about the capabilities of girls and minority students contribute to the problem of low expectations. I teach multivariate statistical analysis in our PhD program at the Claremont Graduate University. Many of my students, especially women and students of color, begin the course with a fear of mathematics and aren't sure they can do it. (Graduate students tend to dread statistics and put it off as long as possible.) Yet almost every student I have taught is capable of understanding and conducting sophisticated statistical analyses, such as hierarchical multiple regression. Often my students' negative impressions of mathematics started with a sexist or racist elementary school teacher, a schoolmate who said "girls can't do math," or poor math scores on standardized tests. Sometimes such experiences traumatize students, who then must be convinced about their real ability. Sheila Tobias has articulated and studied the concept of "math anxiety" and has shown how and why this affliction is particularly prevalent among women.[6]

Many teachers erroneously believe that certain kinds of students cannot do mathematics. When students incorporate those devastating myths into their self-concepts, they often lower their aspirations, thereby shortchanging what they can do with their lives.

The book *The Experts Speak* exposes the incorrect predictions made by "experts" about people who went on to become athletes, politicians, poets, and actors. A movie executive summed up one young man's screen test this way: "Can't act, can't sing, balding, can dance a little."[7] He was writing about Fred Astaire. Clifton Fadiman, in a *New Yorker* review of *Absalom, Absalom* by William Faulkner, called it the "final blowup of what was once a remarkable, if minor, talent."[8] Thirteen years later, Faulkner won the Nobel Prize for Literature.

Eight Solutions for Real Reform

Eight changes are needed to improve STEM education and the science pipeline. We need stronger leadership at all levels of education, as well as rigorous evaluation of new programs to make sure they're working. Our teachers must be well prepared in STEM subjects, and they must be able to teach students effectively. We must hold high standards for students as well as teachers, and mentors must play a pivotal role in inspiring students to pursue their goals. Understanding the true value of a college education is critical to any reform initiative. We must close the achievement gap. Finally, research in American universities must be revitalized to ensure our future as a global economic power.

1. Leadership

Many football fans agree that Vince Lombardi was the greatest football coach of all time. His name is etched on the trophy given to each year's Super Bowl winner. Lombardi was hired in 1959 to coach the dismal Green Bay Packers, a team that had lost 10 of its 12 games the previous season. Many felt the team simply lacked the talent to succeed. By 1968 the team had won the first two Super Bowls.

Similarly, although many schools are failing their students today—with failures attributed to poverty, poor teachers, and the like—a few Lombardi-like leaders have excelled. There are some success stories—for example, the "90-90-90" schools, where 90 percent of the students achieve even though 90 percent of the students are minority students and 90 percent of them live in poverty.

In Connecticut, creative and committed leadership has transformed public education. A 1998 study found that by that year, "Connecticut fourth-grade students ranked first in the nation in reading and mathematics on the National Assessment of Education progress, despite increased student poverty and language diversity in the state's public schools during that decade," and that, in the world, "only top-ranked Singapore outscored Connecticut students in science."[9]

2. Evaluation

Any new reform initiative must be evaluated. Did it work? If so, which components worked best? If not, what can we learn from this failed experiment? Vast sums of money have been spent on innovations that were never evaluated with rigor. We should not change our schools and colleges just because everybody "feels good" about a reform program. True reform must work.

We need both quantitative and qualitative evaluations. Quantitative evaluations of measurable outcomes can provide mathematical models and estimates of program success. Qualitative evaluations explore the nuances of teacher-student interactions during the learning process. Too great a focus on quantitative studies can yield reports with many charts, graphs, and formulae that simply fail to portray what really happens in successful classrooms. Too great a focus on qualitative research can yield glowing reports based on small, biased samples that may not be representative of the larger population. Well-designed reform initiatives—and the proper means of evaluating their efficacy—are critical to improving STEM education in America.

3. Better Teachers

> If we don't step up to the challenge of finding and supporting the best teachers, we'll undermine everything else we are trying to do to improve our schools.
>
> Louis V. Gerstner, Jr., Former Chairman, IBM[10]

To improve science education, we must first improve the selection and education of science teachers. Teachers have the power to inspire students to see science as fun and interesting. John Eichinger, a researcher and former junior high school science teacher, commented that "honest curiosity, fallibility, and enthusiasm are the science teacher's most powerful tools."[11]

Twenty-five years ago, nearly all high school science and math teachers had college degrees in mathematics, physics, chemistry, or biology. Today, teachers are just as likely to have degrees in education as in mathematics and physics. Nearly one-fifth of high school students and over 50 percent of middle school students are enrolled in math classes whose teachers neither majored nor minored in math in college.[12] Palestinian students in the West Bank and students from Lithuania are more likely than American students to have teachers with a math degree.[13]

There clearly has been a decline in the quality of science and mathematics teachers as an unexpected side effect of the women's movement. Happily, increased career opportunities for women since the 1970s have meant that intelligent young women—including those talented in math and the sciences—are no longer constrained to careers in teaching, nursing, and one or two other fields. The result is that the educational system has lost the "hidden subsidy" it enjoyed when so many bright young women chose teaching.

Policymakers and citizens are too complacent about teacher preparation. About 50,000 new teachers start work each year, many at some of the country's neediest schools, without any teacher training.[14] In other countries, teachers are better prepared, and accordingly they earn more money. In chapter 1, I discuss the respect and support accorded teachers in Japan, where "teacher salaries are comparable to those of engineers," according to Linda Darling-Hammond. There, she continues, "the national government eliminates local disparities by providing 50 percent of teachers' salaries to ensure that all are paid at the national level. Thus, shortages are rare, and places in teacher education programs are highly competitive."[15] In contrast, one study that reviewed lessons taught by U.S. teachers found that 87 percent of those lessons were of low quality and none were of high quality: "By comparison, only 13 percent of Japanese and 40 percent of German lessons were judged to be of low quality."[16]

Effective teaching isn't done using a mechanical model. Learning involves active engagement by the learner, who must incorporate the new material. Engaged learning involves linking the new information with one's existing knowledge, based on classroom learning as well as experience and learning outside the classroom. Teachers at both the pre-college and college levels must find ways to connect to the frame of reference in the student's mind. Doing so improves the quality and effectiveness of the instructor's teaching. In my teaching, I use metaphors and stories to link new material with existing knowledge. *The benefits of engaged learning are reciprocal.* Scientific research, theorizing, and model-building are enriched—and may be transformed—if we learn more about the cultural perspectives that underrepresented groups bring to the table.

4. High Expectations

> We only become what we are by the radical and deep-seated refusal of that which others have made of us.
>
> Jean-Paul Sartre[17]

Uri Treisman, a graduate student at Berkeley in the 1980s, conducted some of the most exciting research about *how* people learn mathematics. While a teaching assistant in calculus courses, Treisman observed that the African American students performed very poorly, while the Chinese students excelled.[18] But he did not accept the conventional wisdom that the low achievement rates of the African American students were due to poverty or the underperforming schools they had attended. Instead, he suspected that how they studied, what they studied, and with whom they studied were to blame. So Treisman spent 18 months observing both the Chinese and the African American students. He found that the Chinese spent longer hours studying than other students and that they frequently studied together in groups, for example, working out extra homework problems. Treisman then developed an experimental workshop in which he replicated these interaction and study patterns with the African American students. The results were astounding: the African American students went on to excel in calculus.

Treisman's work, related research by Martin Bonsangue, and work at Harvard by Richard Light have revealed what strategies are successful in college-level mathematics and science instruction.[19] There is evidence that these findings also apply at the pre-college level. Robert Reich has observed that in America's best classrooms,

> instead of individual achievement and competition, the focus is on group learning. Students learn to articulate, clarify, and then restate for one another how they identify and find answers. They learn how to seek and accept criticism from peers, solicit help, and give credit to others. They also learn to negotiate—to explain their own needs, to discern what others need and view things from others' perspectives, and to discover mutually beneficial resolutions. This is an ideal preparation for lifetimes of symbolic-analytic teamwork.[20]

Students learn more effectively when they engage in dialogue with other students and with their instructors. They learn more when they are expected to excel, rather than criticized for their need for remediation. Self-esteem and self-efficacy are central to educational achievement and to the development of meaningful educational and career aspirations. In 1966, the Equality of Educational Opportunity Study (also known as the Coleman study, for the principal investigator, James Coleman) included findings about self-efficacy that foreshadowed these results.[21]

In every educational enterprise we should use a talent development model rather than a model of exclusion.

In another classic study, Robert Rosenthal and Lenore Jacobson took a novel approach to examining how teacher expectations may affect student achievement.[22] They randomly selected students who had not performed particularly well on aptitude tests. Next, the researchers told the students' teachers that, in fact, these young people were potential geniuses whose extraordinary talents were not revealed on intelligence tests. Afterwards, the teachers became more inclined to answer students' questions, to interpret their questions as intelligent, and to spend time working with the students. The students' academic performance improved dramatically. Perhaps the perception that these students were bright affected the subjective component of the teachers' grading process. In any case, these "average" students excelled once their teachers believed that they could.

5. *Committed Mentors and Role Models*

Senator John Glenn once gave a vivid description of how his interest in science was stimulated by a high school teacher who invited Glenn to join him and his family on a summer trip. Senator Glenn described with great enthusiasm how he saw steel being made in Pittsburgh and then visited Niagara Falls, where he watched the generators in awe. He went on to study chemistry at Muskingum College before he enlisted in the military in World War II. We go to great lengths to teach people how to be good teachers. But we must think creatively about how to draw more talented and committed young people into careers as science teachers and as scientists.

At the college level, students planning careers in science need to interact with and learn from professors who are excellent research role models. Unfortunately, students are rarely taught by instructors who are actively engaged in research and can model this process for them. Many undergraduates at large universities are taught by teaching assistants.

The disconnect between the distribution of federal funds and the distribution of talented academic scientists is also to blame for the lack of good role models. This disparity has damaged the productivity of a generation of young researchers.[23] For many years, federal funds for university research have been distributed to relatively few institutions. Each year, about half the federal support for basic scientific research is awarded to 20 universities. However, the data suggest that the top young researchers are not necessarily at those 20 institutions. Over the last three decades a "tenure logjam" has emerged at some leading institutions, with the result that there are very few new job openings in those schools. Consequently, some of the most talented new PhDs take jobs in second- and third-tier institutions. Sometimes the best young physicists from Harvard, Berkeley, and Michigan end up at the University of Arkansas and North Dakota State University.

In one study, my staff and I analyzed surveys of 60,000 scientists and performed hundreds of interviews with scientists and administrators. The data revealed that the continued concentration of federal science funds may be destroying the potential productivity of brilliant young scientists at second- and third-tier universities.[24] But thousands more of our young people are attending schools like Montana State, Kansas State, and the University of Missouri than are attending schools like Harvard, Stanford, and Yale.

6. The Value of a College Education

On the television program *The Apprentice,* host Donald Trump pits teams of contestants against one another to compete for a chance to work for his real estate empire. With a degree from the Wharton School of Business, Trump knows the value of a college education. But he also understands that a college degree does not necessarily translate into success.

One season, a college-educated team ("book smarts") competed against successful businesspeople without college degrees ("street smarts"). Trump suggested that the competition might reveal whether a college degree is needed to succeed in business. In the previous season of the program, contestant Troy was fired partly because he lacked a college education. A year later, Trump told another player, Kevin, that he had too many degrees and too little business experience. In developing a plot line of street smarts versus book smarts, Trump turned his own inconsistency and ambivalence to his advantage.

Who is better equipped for professional life, someone with a solid education or street cred? For better or worse, opinions are affected by TV reality shows. I know of one young woman who cited the early wins by Trump's "street smarts" team to justify her decision several years ago to forgo college.

Having published books about the costs and benefits of a college education, I was fascinated by this TV theme. While the choice of a winner on *The Apprentice* did not provide a definitive answer, it certainly provoked many to debate the issue. People who think college is overrated point to Bill Gates, a college dropout whose net worth exceeds the gross domestic product of most countries. But decisions should be based on all the data, not the dramatic exceptions. Otherwise, it's like gambling your future in a Trump casino. The odds are against you.

7. Closing the Achievement Gaps

Competent mathematics and science education is critical if the United States is to compete in a globalized, high-tech information economy. International assessments of educational achievement in these subjects consistently place U.S. high school students at or near the bottom of all industrialized countries. Furthermore, high-stakes tests also reveal a disturbing achievement gap between white students and students of color within the United States.

For the past nine years, Martin Bonsangue and I have worked with a consortium of Houston's colleges and universities to increase achievement levels and close the racial gap. The institutions—the University of Houston, Texas Southern University, the University of Houston Downtown, Texas State University, Rice University, Houston Community College, the University of Houston at Victoria, and San Jacinto Community College—have been supported in these efforts by a National Science Foundation Louis Stokes Alliance for Minority Participation (LSAMP) grant focused on STEM disciplines.

During the first five years of the funding (Phase I of the grant), the numbers of minority bachelor's degrees earned in STEM disciplines increased by 15 to 20 percent per year, falling just short of doubling in five years. This Houston growth rate was much greater than the national growth rate for minority students. In chapter 6 I discuss how this turnaround was achieved, as well as some of the extraordinary students who benefited.

8. Revitalizing University Research

Controversies about federal science funds for university research have become highly visible—and acrimonious. Debate about who should get support and why, long the province of a carefully developed system of peer, or merit, review, has been the subject of special task force reports, congressional hearings, and cover stories.

A growing number of second- and third-tier universities, feeling iced out of the competition, hire lobbying firms, companies that pride themselves in assisting universities to circumvent peer review by getting facilities and other awards listed as line items in otherwise unrelated congressional legislation.

Rhetoric has become overheated, ranging from allegations that pork-barrel science will eradicate merit to claims that only a few universities in this country have scientists qualified to conduct cutting-edge research.

This is not an isolated academic tea-party discussion about who gets how many crumbs from the funding pie (although it has been suggested that academic politics get unusually vicious because the stakes are so low). The consequence of this debate will be of fundamental importance for the technological strength of our economy.

Most science-policy analysts agree that the chain of technological innovation goes something like this: Federal funding leads to basic research which leads to applied research which leads to development, production, and dissemination or marketing. This is, admittedly, a simplified version of the process. But most technological innovations begin with basic research, and most of the nation's basic research is done in our universities. Federal science funding policy, then, is an issue that cuts to the heart of our competitive stance as a nation.

The controversies swirl around an assumption and a fact. The assumption is that the only basis for awarding federal science funds is merit. Simply put, we want to fund the best scientists doing the most creative projects. Most policy-makers and analysts agree that this is the only defensible policy. And we believe that there is no justification for end runs around peer review, like those executed by lobbyists. If a new medication turns out to have devastating side effects, we don't want it to be because second-rate researchers got funded under pork-barrel rules. Nor do we want it to be because a second-rate researcher at one of the country's leading research institutions beat out a better researcher with more creative ideas who happened to be employed at a rural state university. Evidence indicates that the latter threat is more likely than the former. The practical implications of early superconductivity research were judged by experts to be staggering. The scientist who made a major discovery was employed by the University of Alabama, Huntsville, not one of the top 20 research universities.

That's the assumption. The fact is that a huge proportion of federal science funds go to the leading institutions. About half of such funds are awarded to the leading 20 universities each year. This has been true for decades. The struggle about precisely this issue delayed creation of the National Science Foundation, the principal agency funding university research.

Proponents of the present concentration of funds argue that NSF and other federal agencies must base decisions solely on the excellence of proposed research. If the best researchers—and the best ideas—happen to be in a few leading institutions, then there is no question where science funding should be directed.

Proponents of a greater geographical dispersion of funds argue that potentially creative scientists are not being trained or nurtured in sparsely developed regions, that there are significant benefits to a local region from having centers of scientific excellence, that progress in basic research depends to some degree upon a synergistic reaction with industrial R&D, and that some industries thrive in regions that do not have a leading university. Good science, they insist, may be overlooked under current funding procedures.

Research has suggested some specific ways to improve the system. These include recommendations for strengthening peer review. Merit review, as in Churchill's description of democracy, may be imperfect, but it is better than any other system available.

All science funding, federal or otherwise, represents risk capital. No investment in research—basic or applied—is guaranteed to yield results. Yet present funding policies are tantamount to assuming that the horse first out of the gate is bound to win the race.

Our funding system has been responding too slowly to shifts in the distribution of scientific talent. In that sense, any argument that forces a painful but necessary reexamination of the funding process is healthy.

The Curriculum: A Deliberate Omission

I don't believe that simply changing the curriculum (yet again) will solve all our educational problems. Yes, we want a current, innovative curriculum, but we have invested way too much time, money, and effort in curriculum reform in the past.

Educators and policymakers engage in fiery debate about the curriculum. Do children learn to read best through "whole language instruction" or phonics? Is using higher-level reasoning and theory the best way to teach mathematics? Or should students go back to basics and learn fundamental arithmetic skills?

Each side sometimes labels and denigrates the other. Whole language and mathematical theory proponents are reviled as weak and permissive. For their part, the whole language and mathematical theory people consider the "back to basics" group to be concrete thinkers incapable of abstract thought or, worse still, hegemonic oppressors.

And so it goes. In most of these curriculum debates, common sense, experience, and research suggest that students will benefit from a combination of approaches. Nobel laureate Niels Bohr once said, "The opposite of a true statement is a false statement. But the opposite of a profoundly true statement may be another profoundly true statement."[25]

I believe good teachers are more important than good curricula. I'm hopeful that in the future, we will have both. However, I would rather see a young person taught by an exciting, engaged, supportive teacher working with an outdated 1950s curriculum than by a boring, hostile, condescending teacher working with the latest curriculum and standards. Students with good teachers tend to be better educated, more interested in math and science, and more successful in their careers.

A Way Forward

Current mathematics and science education programs are not doing a good job of educating students and preparing them for a global economy, but we can turn this situation around. By providing incentives to recruit outstanding young people into teaching, encouraging professional development, and raising expectations, we can substantially improve America's teaching force.

We also must improve students' participation in STEM fields. The era of restricting access to a solid education has passed. We must open access to math and science as soon as possible.

An informed citizenry capable of making sound judgments also will be needed to set the boundaries for science so that we can continue to reap its rewards. Scientific research and development cannot be conducted independent of the society that will interpret and apply the findings, just as Heisenberg discovered that the behavior of subatomic particles cannot be described in isolation, independent of the observer. Our scientific theories, hypotheses, and interpretations—and indeed our nation's place in the world—will be strengthened if the research is conducted and informed by people with differing worldviews and cultural perspectives. Science has contributed so much to the quality of our lives and promises to contribute even more in the near future.

Throughout this book I will describe what is currently holding us back, but also how we can find a way forward. We *can* reform education, produce higher-achieving students, and renew America's standing as a leader in science, technology, engineering, and mathematics.

CHAPTER 1

America's Place in the World

> If an unfriendly foreign power had attempted to impose on America the mediocre educational performance that exists today, we might well have viewed it as an act of war.
>
> *A Nation at Risk*

The National Commission on Excellence in Education declared an emergency in American education a quarter century ago. Has the mediocrity been replaced by excellence? The short answer is no. In 2007, in another National Commission report, Intel Corporation spokesman Howard High soberly predicted even more outsourcing, seeking talent overseas: "We go where the smart people are. Now our business operations are two-thirds in the U.S. and one-third overseas. But that ratio will flip over the next ten years."[1]

Technology has changed the world of work, with the quality of STEM education driving economic productivity. Study after study has shown that U.S. students lag behind their foreign counterparts. We have been trying to prepare young people for a 21st-century workplace with 19th-century educational structures.

"Our Quiet Crisis"

Rapid advances in communications technology have reshaped how business and commerce function. Thomas Friedman reported these advances in his book *The World Is Flat*. Friedman identifies forces that leveled the international playing field in business, including the end of the Cold War; the expansion of the World Wide Web; software that streamlined workflow; job outsourcing; open-sourcing of software development; improvement of supply chains; powerful Internet search engines, especially Google; and the use of cell phones and PDAs to link to global electronic networks.

Friedman describes a world in which economic production and distribution are conducted by global collaborators employing the latest technology at warp speed. This is the advice he gives his daughters: "Girls, when I was growing up, my parents used to say to me, 'Tom, finish your dinner—people in China and India are starving.' My advice to you is: Girls, finish your homework—people in China and India are starving for your jobs."[2]

Friedman believes America's economic future is linked directly to the quality of our educational systems: "Because it takes fifteen years to create a scientist or advanced engineer, starting from when that young man or woman first gets hooked on science and math in elementary school, we should be embarking on an all-hands-on-deck, no-holds-barred, no-budget-too-large crash program for science and engineering education immediately. The fact that we are not doing so is our quiet crisis."[3]

Policymakers and economists disagree about which nations will boast the world's most powerful economies in the near future. Economist Lester Thurow analyzed the economic potential of the United States, Europe, and Japan in his book *Head to Head*. He linked future economic success to education. He noted that while workers on an assembly line in the manufacturing economies did not need to apply deep analytical reasoning, workers at every level in a high-tech information economy will need such analytical skills. "If the bottom 50 percent [of workers] cannot learn what must be learned, new high-tech processes cannot be employed."[4]

In 1973, at the height of the Cold War, and in 1998, a decade after the Berlin Wall fell, I presented papers at conferences in Budapest. The changes in the city's energy and vitality were palpable. The latter conference focused on the emerging economies of Eastern Europe. We saw great potential for growth in those former Soviet satellites. A 2005 article about Eastern European technological growth noted that those economies indeed have grown and prospered: "Today the region is sucking in foreign investment at a rate of $37 billion annually, which places it second to China in the international competition of capital and light years ahead of India. Central European stock markets are taking off, too."[5]

Western Europe is emerging as an economic force. However, some feel that demographic forces may doom the European Union. An optimist about Europe notes that "the new United States of Europe—to use Winston Churchill's phrase—has more people, more wealth, and more trade than the United States of America."[6] A pessimist observed, "The European Union is historically nuts. It does not reflect the will of a single nation-state, or the will of an empire, based on the ability of a central political entity to dominate its periphery, or some form of established European national identity with deep historic roots."[7] The nations of Southeast Asia have also been considering an economic confederation. In

2005, leaders of 10 nations from that region met with their counterparts from other Asian and Pacific powers, including Japan, India, China, and Australia.

And where does this leave the United States? According to a National Academy of Sciences report, "chemical companies closed 70 facilities in the United States in 2004 and tagged 40 more for shutdown. Of 120 chemical plants being built around the world with price tags of $1 billion or more, one is in the United States and 50 are in China. No new refineries have been built in the United States since 1976."[8] To illustrate these trends, James Johnson and John Kasarda provide the following scenario:

> Joe Smith started his day early, having set his alarm clock (*made in Japan*) for 6 A.M. While his coffee pot (*made in China*) was perking, he shaved with his electric razor (*made in Hong Kong*). He put on a dress shirt (*made in Sri Lanka*), jeans (*made in Singapore*), and tennis shoes (*made in Korea*).
>
> After cooking his breakfast in his new electric skillet (*made in India*), he sat down with his calculator (*made in Mexico*) to see how much he could spend today. After setting his watch (*made in Taiwan*) to the radio (*made in India*) he got in his car (*made in Japan*) and continued to search for a good paying American job.
>
> At the end of yet another discouraging and fruitless day, Joe decided to relax for a while. He put on his sandals (*made in Brazil*), poured himself a glass of wine (*made in France*) and turned on his TV (*made in Indonesia*), and then wondered why he can't find a good paying job in . . . *AMERICA*.[9]

Experts may disagree about which nations will have the strongest high-tech economies, but STEM education is essential in this new "flat world."

Education and Economic Productivity

The nature of commerce has changed. The United States is not as competitive as it was only a quarter century ago. The bellwether indicator of the shifting economic fortunes of the United States has been the automobile industry.

In *The Reckoning*, David Halberstam described the decades over which the American automobile industry faltered while the Japanese automobile industry soared. Halberstam's search for clues and explanations focused on education. He observed that the Japanese had a national commitment to economic success. "The dream of Japanese greatness through military power had proved a false and destructive one. The ashes of that dream were all around them. There had to be another path. Quickly a consensus evolved: Japan was so limited in size and natural resources, so vulnerable—as seen at Hiroshima and Nagasaki—to modern weaponry, that it could become strong only if it focused all of its energies on commerce and completely avoided military solutions."[10]

One of the most significant decisions made by the Japanese establishment was to strengthen education generally and engineering education in particular. Halberstam quotes historian Frank Gibney, who stated that Japan's educational system was "the key that winds the watch" (276). In Japanese society, social mobility was encouraged, the objective being a well-educated meritocracy. In addition, teachers were afforded special social status:

> The educational system was critical to the rise of modern Japan: it crystallized, legitimized, and modernized values already existing. It removed a great deal of potential class resentment on the part of the poor, *and it provided Japan with an extraordinarily well qualified, proud, amenable, and ambitious working class.* If modern post-war Japan was perhaps the world's most efficient distribution system, in which remarkably little was wasted, in human, material, or capital terms, then a vital part of it was the educational system: it supplied the nation with the right number of workers, the right number of engineers, and the right number of managers for every need of a modern society. *But, equally important, it brought to the poorest homes the sense that there were better possibilities for the children.* (276–77; emphasis added)

Psychological testing performed during the 1950s revealed that the Japanese viewed education as vital to success. These values were similar to those that immigrants to America had passed on to their children, but Americans themselves, now accustomed to affluence, no longer stressed the importance of education (277). Japan also expanded engineering education and scholarships. The Japanese elite hoped to flood factories with well-trained, capable engineers. "They might work on something tiny, perhaps a task a good deal smaller than what they had dreamed of while they were students," says Halberstam, "but the cumulative effect of so many talented engineers working on so many small things would be incalculable" (278).

Contrast Japan's educational transformation with an episode from American colonial history. Thomas Jefferson believed that science and mathematics education was crucial to the development of a new nation. Jefferson invited the French minister of agriculture, Pierre S. du Pont de Nemours, to review education in the United States. Du Pont agreed that science and mathematics were important and lamented that there were no textbooks in the United States on these subjects. Jefferson asked Congress for $10,000 to write science textbooks for each grade level that related science and mathematics to the welfare of the nation. Congress turned down his request, saying that jurisdiction over church and schools belonged to the local communities.

Writing about the historical and philosophical foundations of science literacy, Paul DeHart Hurd noted: "In terms of relating science and technology to human affairs, and social and economic progress, we are about at the same place we were two hundred years ago or even 400 years ago. If we had been successful in relating science to the welfare of the individual and the common good, there would have been no need for the report called *A Nation at Risk* and the 300 national reports that have emerged since."[11]

For a period following World War II, most Americans viewed their country as world leaders in just about everything. The United States had led the free world to victory over the Axis powers. The development of nuclear technology yielded both an awesome weapon and groundbreaking technology with peaceful applications. The U.S. economy was a dominant force on the international scene. Americans had the best universities. It seemed that the rest of the world wanted to copy American culture: the Japanese loved baseball, and the Soviets claimed that they had invented it. Marilyn Monroe and James Dean had fans in Thailand, Australia, Argentina, and France.

Times have changed. Most Americans now question whether the United States can compete economically with China, Germany, and Japan. Virtually everyone in America believes that Japan makes the best cars, a particularly disturbing realization, since the automobile industry was once the heart of the American economy. Less than 60 years after Eisenhower cabinet member Charles E. Wilson declared, "What's good for General Motors is good for the United States," General Motors declared bankruptcy.

Our postwar view of America's preeminence probably was too rosy. And our current fears about America's decline and economic vulnerability are almost certainly too pessimistic. But STEM education is critical for strengthening our economy.

"Our People Can't Hack It"

Substantial numbers of the graduate students in science, technology, engineering, and mathematics in the United States are international students.[12] According to Cal Tech's David Goodstein:

> [Our] engines of research are fueled by foreign graduate students, from around the world, and particularly from the Pacific Rim. The role that Greece once served for the Roman Empire, and that Europe served for prewar America, America is now serving for our friends from across the ocean. We are the source of culture and learning, at least in physics. Historians may or may not ultimately conclude that this was the era in which the torch of world leadership was passed from us to other hands. But there is no denying the fact that, at least in physics education, we are now playing the role traditionally reserved for the world leader of the previous era.[13]

I do not believe that the presence of international students in our PhD programs is a problem. I agree with Sheldon Glashow, a Harvard Nobel Prize winner in physics, who said more than 20 years ago: "We import them. We import them from Iran; we import them from Turkey, and Taiwan, from whatever countries we can find. They stay in this country—more than half of them stay in this country—they take all the good jobs. . . . This is not a problem. This is the solution. These new Americans who are a vital part of America today and are the technological backbone of the country are the solution. Our people can't hack it."[14]

For many decades, naturalized citizens have been among the top scientists in America. The first influx of highly productive scientists and engineers from foreign lands were the intellectual elites who escaped Nazi Germany. The current influx of graduate students, including recent immigrants from Southeast Asia, may become the technological backbone of our nation's scientific and industrial effort. In fact, two-thirds of the science and engineering PhD students in the United States are from abroad. But American science education will be compromised if our national research enterprise becomes dependent upon immigrant talent, particularly in the post-9/11 era.

New immigration laws may discourage foreign scientists from remaining in our country after completing their education. Post-9/11 immigration policies have dramatically reduced the number of foreign nationals with technical expertise permitted to work in the United States. George Will points out, for example, that our current policy includes "the compulsory expulsion or exclusion of talents crucial to the creativity of the semiconductor industry that powers the thriving portion of our bifurcated economy."[15] The long-term implications for national research and development could be devastating.

In 2005, a distinguished panel of scientists and business leaders commissioned by the National Academy of Sciences produced the report *Rising above the*

ARE YOU SMARTER THAN A FOURTH GRADER?

On the television quiz show *Are You Smarter Than a Fifth Grader?* adults are asked fifth-grade questions for a chance to win a million dollars. See how well you fare on the National Assessment of Educational Progress (NAEP) tests, which are given to fourth and eighth graders in all 50 states plus the District of Columbia. (The answers are in the box footnote.)

First, a fourth-grade algebra question:

N stands for the number of hours of sleep Ken gets each night. Which of the following represents the number of hours of sleep Ken gets in 1 week?

a. $N+7$ b. $N-7$ c. $N\times 7$ d. $N\div 7$

(This question was answered correctly by 61 percent of fourth graders in the United States.)

Here is an eighth-grade algebra question:

1, 9, 25, 49, 81, . . .

The same rule is applied to each number in the pattern above. What is the sixth number in the pattern?

a. 40 b. 100 c. 121 d. 144 e. 169

(This question was answered correctly by 60 percent of eighth graders in the United States.)

Source: National Assessment of Educational Progress. *The Nation's Report Card: Mathematics 2005*, National Center for Education Statistics (Washington, DC: U.S. Department of Education, Institute of Education Sciences, 2006). Percentages of students with correct answers, p. 31.

Answers: c, c.

Gathering Storm: Energizing and Employing America for a Brighter Economic Future. Among their key recommendations:

> Make the United States the most attractive setting in which to study and perform research so that we can develop, recruit, and retain the best and brightest students, scientists, and engineers from within the United States and throughout the world. . . .
>
> . . . Provide a one-year automatic visa extension to international students who receive doctorates or the equivalent in science, technology, engineering, mathematics, or other fields of national need at qualified U.S. institutions to remain in the United States to seek employment.[16]

What is most disturbing is Professor Glashow's observation, above, that "our people can't hack it." He's right: "While we are asleep tonight," according to testimony by Margaret Spellings before the Congressional Committee on Education and the Workforce, "accountants in India will do our taxes, radiologists in Australia will read our CAT scans and technicians in China will build our computers. As other nations race to catch up, there is mounting evidence that American students are falling behind. I know you know the numbers, but they bear repeating. Currently, our 15-year-olds rank 24th out of 29 developed nations in math, literacy and problem solving. Half of our 17-year-olds don't have the necessary math skills to work as a production associate in a modern auto plant."[17]

Can American Students Compete on the International Stage?

Frank Layden, former coach of the Utah Jazz, experienced great frustration with one player's inability to learn a new basketball maneuver despite extensive practice. Finally, he said to the player: "We've been working on this every night for weeks. Yet you still can't seem to get it. Frankly, I can't decide if your problem is ignorance or apathy." The player glared at him and replied, "I don't know and I don't care."

Our young people don't know enough about mathematics and science, and most of them don't care. As students progress through the school system, they tend to find science less interesting. R. E. Yager and J. E. Penick presented data as early as 1986 showing this decline in interest over time and concluded that "the more years our students enroll in science courses, the less they like it. Obviously, if one of our goals is for students to enjoy science and feel successful at it, we should quit teaching science in third grade. Or perhaps we should try teaching it differently."[18]

So, where do we stand internationally on science and mathematics education? A false syllogism has driven much of the debate about educational policy in recent years. It goes something like this: "Our high school students perform at or close to the bottom in international assessments of educational achievement. This shows how terribly American schooling deteriorated in the past half century. Why did this happen? I believe the reason is . . ." Then the speaker will invoke his or her pet explanation for the decline in America's educational system.

This is false logic based on erroneous assumptions. Yes, our high school students perform poorly on international assessments. But, no, this does not

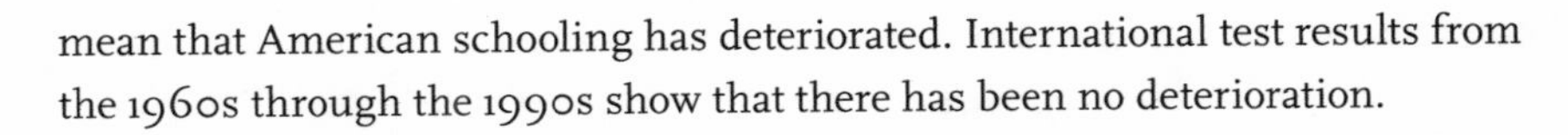
mean that American schooling has deteriorated. International test results from the 1960s through the 1990s show that there has been no deterioration.

American high school students have *always* performed at or near the bottom in international assessments.

International Assessments in the Twenty-First Century

Since 2000, the United States has participated in five international surveys:[19]

1. Program for International Student Assessment (PISA): Reading, Mathematics, and Science, conducted by the Organisation for Economic Co-operation and Development (OECD): 15-year-olds
2. Adult Literacy and Life Skills Survey (ALL): 16- to 65-year-olds
3. Trends in International Mathematics and Science Study (TIMSS): grades 4 and 8, in both 2003 and 2007
4. Progress in International Literacy Study (PIRLS): grade 4

In addition, one recurring study is limited to the United States, the National Assessment of Educational Progress (NAEP), also known as "The Nation's Report Card." In this study, students from each state, mainly in grades 4 and 8, are tested.

The following paragraphs examine the findings of the PISA study conducted by the OECD and the TIMSS surveys because these two assessments provide a good snapshot of students at two different developmental levels, testing for different sets of skills. Robyn Baker and Alister Jones draw this distinction between the TIMSS and PISA studies: "While the TIMSS is designed to link with the current school science curriculum, the PISA is based on quite a different premise. The PISA seeks to assess what students might require in the future and so its focus is on whether young adults, at age 15 years, have the ability to use their knowledge and skills to meet real-life challenges, rather than whether they have mastered a specific school curriculum."[20]

How have American students performed on the OECD and TIMSS assessments?

U.S. Rankings in OECD Data

Data from the OECD 2003 PISA international assessment show that of 29 OECD countries studied, 23 countries produced higher mathematical literacy scores among 15-year-olds than did the United States (table 1.1). Similarly, of the 29

TABLE 1.1
Ranking of countries in the OECD math literacy study, 2003: Math literacy among 15-year-olds

1. Finland	11. Iceland	21. Poland
2. South Korea	12. Denmark	22. Hungary
3. The Netherlands	13. France	23. Spain
4. Japan	14. Sweden	**24. United States** / (Latvia)[a]
5. Canada	15. Austria	25. Portugal
6. Belgium	16. Germany	26. Italy
7. Switzerland	17. Ireland	27. Greece
8. Australia	18. Slovak Republic	28. Turkey
9. New Zealand	19. Norway	29. Mexico
10. Czech Republic	20. Luxemburg	

Source: Mariann Lemke and Patrick Gonzales, *U.S. Student and Adult Performance on International Assessments of Educational Achievement: Findings from the Condition of Education 2006*, NCES 2006-073 (Washington, DC: U.S. Department of Education, National Center for Education Statistics, June 2006), p. 16.

[a]Students in Latvia, which is not an OECD country, achieved the same score as U.S. students.

TABLE 1.2
Ranking of countries in the OECD science literacy study, 2003: Science literacy among 15-year-olds

1. Finland	11. Belgium	21. Spain
2. Japan	12. Sweden	22. Italy
3. South Korea	13. Ireland	23. Norway
4. Australia	14. Hungary	24. Luxembourg
5. The Netherlands	15. Germany	25. Greece
6. Czech Republic	16. Poland	26. Denmark
7. New Zealand	17. Slovak Republic	27. Portugal
8. Canada	18. Iceland	28. Turkey
9. Switzerland	**19. United States**	29. Mexico
10. France	20. Austria	

Source: Lemke and Gonzales, *U.S. Student and Adult Performance on International Assessments*, p. 22.

countries studied, 18 ranked above the United States in the OECD science literacy assessment of 15-year-olds (table 1.2).

Just what does this all mean? Consider a sports analogy: If we compare U.S. math and science scores with 2010 NFL football rankings, the United States would be the Cleveland Browns. If we compare them with 2010 NBA basketball rankings, the United States would be the Golden State Warriors. Each of these teams played below .500, but each had a credible record. Neither team occupied last place in the league, yet neither team made the playoffs, either.

U.S. Rankings TIMSS Data

TIMSS surveys focus on the fourth and eighth grades. From 1995 to 2003 there was no change in the mathematics performance of American fourth graders.

But over the same time period, the performance of eighth graders did improve: compared with their international peers they moved up from 13th place in 1995 to 8th place in 2003.[21]

Overall, boys outperformed girls in mathematics in both grades. However, between 1995 and 2003, the achievement gap narrowed between white and black students in the fourth and eighth grades.[22] Improvement in the achievement gap continued between 2003 and 2007. The 2007 study found that only Taipei, Republic of Korea, Singapore, Hong Kong, Japan, Hungary, England, and the Russian Federation outperformed the United States in mathematics at the eighth grade level. American eighth graders outperformed their counterparts from 36 other nations. Nevertheless, the TIMSS researchers observe, "When older U.S. students are asked to apply what they have learned in mathematics, they demonstrate less ability than most of their peers in other highly industrialized countries."[23]

In summary, while American eighth graders have improved, these 21st-century international comparisons consistently reveal weak performance by American high school students.

It is important to place these 21st-century international assessments of education achievement in a historical context. I have systematically analyzed the results of prior international assessments from the 1960s through the 1990s. Those studies are presented in the appendix. Readers who want to trace the performance of American students over the past 50 years or who want to study the

TABLE 1.3

Performance of American high school students in international mathematics and science achievement assessments

Year	No. of countries in study[a]	U.S. rank
Math assessments		
1965	12	12
1989	12	12
1991	15	12
2003	29	24
Science assessments		
1973	14	14
1988 (biology)	13	13
1988 (chemistry)	13	9
1991 (physics)	15	13
2003	29	19

[a]For consistency, I have included in each count developing nations (i.e., China, India, Iran, and Thailand) that previously had been omitted from a prior tabulation but were later included in others.

specific findings from a given international assessment should turn to the appendix. I summarize the lessons learned from these many studies in the following, concluding section of this chapter.

In Closing

An interesting fact emerges from the analyses conducted to estimate the effects of differing national school attendance rates. While the United States has a very high percentage of 18-year-olds in school relative to other countries, *the percentage of those in U.S. schools who study mathematics and science is very low*! The statistics on the relative percentage of American students who study advanced mathematics—for example, calculus—are even more disturbing. Keep these findings in mind as we explore what works in American STEM education in succeeding chapters.

Table 1.3 summarizes the results of 40 years of international assessments of math and science education. For each assessment, I indicate the number of countries surveyed and where the United States ranked. Three observations can be drawn from these international data:

1. At no time was the performance of U.S. students excellent, or even average.
2. In contrast with the conventional wisdom that U.S. performance has declined in recent decades, performance has actually improved. The hard work of American teachers, students, and parents has started to pay off.
3. The findings are mixed over a 40-year period, but there is some evidence that American students have done well on items that measure advanced analytical reasoning.

In my research I have interviewed hundreds of undergraduate STEM majors. I have encountered scores of students who received part of their precollege education in a foreign country and part in the United States. I always ask them which country had the toughest, most demanding curriculum? Where did they learn the most? With only one exception, these students have said that their foreign schooling was far more advanced than their U.S. schooling.

One undergraduate transferred to a small-town high school in Texas after completing two years of high school in the Philippines. He said, "In Manila, I was in danger of flunking out. In Texas, I became the valedictorian."

CHAPTER 2

The Achievement Gap

> Few tragedies can be more extensive than the stunting of life, few injustices deeper than the denial of an opportunity to strive or even to hope, by a limit imposed from without, but falsely identified as lying within.
>
> Stephen Jay Gould

Let's recap what we have learned so far:

1. In a high-tech global economy, STEM education assumes more importance than ever before.
2. American high school students perform poorly in STEM fields in comparison with their counterparts from other nations.
3. America's low rank among other countries has persisted for 40 years.
4. That Americans value—and are committed to—mass public education, while many other nations educate only an elite group of students does not explain away the findings.
5. The percentage of American adolescents who study mathematics and science is lower than that of other countries.

Why are we not competitive in STEM education? In preparing a previous book, *Aptitude Revisited: Rethinking Math and Science Education for America's Next Century,* I concluded that there was one fundamental flaw in our system. Other factors, especially poverty, played a role, but there was, and still is, one basic problem: *Far too many students are blocked from opportunities to master STEM because of false assumptions about aptitude.*

The barriers to STEM education take many forms. Students may be dissuaded from taking advanced mathematics and science courses. A teacher may interpret a probing question as a stupid one. A counselor may recommend that a student discontinue further study.

Richard Tapia is a world-class, award-winning professor of mathematics at Rice University and a member of the National Science Board. A Latino from a poor neighborhood in Los Angeles, he attended college at UCLA. When he told his advisor he was thinking of applying to graduate school, the advisor said, "I don't think that's a good idea." Tapia replied, "No problem. You won't be the one going to grad school, I will."[1]

Tapia, one of only six "university professors" in the history of Rice University, has said: "As minority faculty we serve as role models in two directions. We demonstrate feasibility to the minority students and show the non-minorities that we as minorities can be excellent teachers and faculty."[2]

Beliefs about innate aptitude and intelligence have done more harm than good. They are epistemologically unsound, and proponents of such beliefs cannot offer any compelling evidence that they describe empirical reality. Modern-day psychometricians, including Charles Spearman, Robert Sternberg, and Howard Gardner, sound like medieval theologians arguing about how many angels would fit on the head of a pin. Is there one kind of intelligence (Spearman), or are there three (Sternberg), or eight (Gardner)?

Some scientific concepts, hypotheses, and theories that were once widely accepted have been abandoned. We no longer believe that the earth is the center of the universe. We no longer try to understand personality by studying the bumps on someone's scalp (called phrenology). We no longer treat diseases with leeches. It is time to abandon the concept of intelligence or aptitude.

International research about educational achievement has found that when American students do poorly, parents and teachers often blame their underachievement on aptitude. When students in some other countries do poorly, parents and teachers instead say that students aren't working hard enough.

In February of 2001, I wrote about this issue in an article that appeared in *USA Today*. The editors placed a headline on the piece that they thought captured the message: "Tell Students: Yes, You Can." In the article I discussed programs that work to close the achievement gaps between rich and poor students, between majority and minority students, and between girls and boys.

When I wrote *Aptitude Revisited*, I was optimistic about our capacity to turn this situation around. I remain optimistic. If assumptions about group aptitude are arbitrary, a group that now is down can rise up. As you will see below, this has happened to some degree for young women in high school and college. But women still have a long way to go before we have equity in STEM fields.

Someday driver's ed and STEM education might have something in common. High schoolers know that with turning 16 comes the possibility of earning a driver's license and the freedom to drive. Before they may do so, however, they must take a written test, perhaps several times, and a driving test, perhaps several times. But most 16-year-olds are confident that they will pass these tests and become licensed drivers.

This is how we should view STEM education.

In this chapter, I present data that demonstrate what we all know: There are achievement gaps in the United States between young men and women, and between white students and students of color. I also analyze the concept of aptitude and its contribution to these gaps. In chapter 6, I show exactly how we can close these gaps.

STEM education is centrally linked to the future career achievements, and ultimately, the power, of young people. Unfortunately, the people in our society with the least power—the poor, people of color, and women—are the least likely to study mathematics and science. In the postindustrial age, science and technology have assumed center stage. Those who fail to obtain an adequate background in these subjects in elementary school, junior high school, and high school will be at a competitive disadvantage in higher education and in the work world.

Our current expectations about who can successfully study science and mathematics are based on erroneous assumptions and are totally at variance with the educational research literature. Until we correct these misapprehensions about who can succeed in mathematics and science and begin educating *all* young people, the gap between the haves and the have-nots in our society will continue to widen.

Our schools are not doing a good enough job of educating students, even affluent white male students. If there are any doubts about this proposition, a review of the performance of American youngsters, even the top 1 percent, in international assessments of mathematics and science achievement will quickly dispel them. As demonstrated in chapter 1, American schools have *not* deteriorated during the past 25 years. In fact, the problems have existed for a long time. The resources available to solve these problems include dedicated, hard-working, capable teachers who labor mightily every day to deliver a quality education to their students; parents who want the best for their children; and most importantly, talented students themselves.

Gender Gaps

Studying mathematics can be critical to the development of a successful career, and yet research shows that women study mathematics less than men do. The gender gap in elementary and secondary school has shifted in the past decade. One analysis identified specific trends favoring girls. According to Deborah

Taylor and Maureen Lorimer, boys score lower in language arts tests, are far more likely to be in special education, are more likely to be labeled as learning disabled, are disciplined more often, enroll in fewer advanced courses, and are less likely to go to college. Three-fourths of the nearly one million children taking the prescribed attention-deficit drug Ritalin are boys.[3]

Similarly, the gender ratio in college has shifted. There are now fewer men than women enrolled in college; the percentage of men declined from 58 percent in 1968 to 44 percent 30 years later.[4] Among traditional-age college students (ages 18 to 25) the male percentage is slightly higher, but among students over 25 there are almost twice as many women as men.[5] However, women are still underrepresented in STEM careers and on college and university STEM faculties. There still is a glass ceiling in academia. The lack of women at the highest levels of achievement may be attributed to the avoidance of mathematics at an earlier stage of schooling.

A recent National Science Foundation report commissioned by the U.S. Congress showed marginal improvements for women in academic STEM disciplines, but at levels far below gender parity. For example, women represented 32 percent of new PhDs in chemistry from 1999 to 2003. In 2003, only 9.7 percent of full professors in mathematics were women. Here are three key findings from this report:

> Although women represent an increasing share of science, engineering, and mathematics faculty, they continue to be underrepresented in many of these disciplines.
>
> Women account for about 17 percent of applications for both tenure-track and tenured positions in the departments surveyed.
>
> There is little evidence overall that men and women spent different proportions of their time on teaching, research, and service.[6]

The report quotes Rhonda Hughes, chair of the mathematics department at Bryn Mawr College, who once said, “Mathematicians are decent types, but sadly, mathematics is still very much a man’s world.”[7]

Helen Astin studied the research experiences of men and women whose work is frequently cited. After comparing highly cited essays written by both male and female authors, Astin concluded:

> When we examine their perceptions about what led them to undertake the highly cited research in the first place, women appear to be responding to others rather than being driven by their own quest. That is, they are less likely to

> undertake the work because they are interested in solving a problem, but rather that the work was the outcome of the dissertation or they were invited to prepare the piece. Furthermore, when it comes to explaining why their work is so frequently cited, the women appear to be more interested in how their work can be useful to others (their research can help and the findings can be applied by others). They also make more positive attributions about the importance of their work than do the men. They see their research as integrating knowledge and providing direction for further work: "a useful procedure for calculating the affinity of the drugs for the receptor"; "the hope that this approach might lead to a new type of cancer immunotherapy."[8]

There were strong differences between the research experiences of social scientists and natural scientists. In fact, according to Astin, "the results reported thus far suggest that overall, field may be more of a factor than gender in the experiences reported by scientists who produce highly cited research."[9]

Ben A. Barres is a neurobiologist at Stanford. A transgendered person, Barres has written about the debate over whether women are innately less able than men in mathematics and science. He says: "If innate intellectual abilities are not to blame for women's slow advance in science careers, than what is? The foremost factor, I believe, is the societal assumption that women are innately less able than men. . . . Shortly after I changed sex, a faculty member was heard to say, 'Ben Barres gave a great seminar today, but then his work is much better than his sister's.' "[10]

Maria Klawe, Telle Whitney, and Caroline Simard reviewed and presented the available data on women in the field of computer science. They report that only 16.2 percent of computer science faculty in four-year institutions are women, and only 12 percent of the full professors are women. African American faculty and Hispanic faculty (including both men and women) represent about 1 to 2 percent. Given equal qualifications, women faculty in computer science earn just 81 percent of what their male counterparts earn. Why is this important? This underrepresentation puts the computing field at a disadvantage: "Diversity leads to better decision outcomes, enhanced task performance, and greater innovation and creativity. The pervasiveness of unconscious biases and stereotyping having to do with the gender and ethnic composition of our technical talent limits the possibilities of technological innovation around the world."[11]

In 2006, a special commission of the National Academy of Sciences released its report on the role of women in academic science and engineering. Their work yielded eight findings:

1. Women have the ability and drive to succeed in science and engineering.
2. Women who are interested in science and engineering careers are lost at every educational transition.
3. The problem is simply not the pipeline.
4. Women are very likely to face discrimination in every field of science and engineering.
5. A substantial body of evidence establishes that most people—men and women—hold implicit biases about aptitude.
6. Evaluation criteria contain arbitrary and subjective components that disadvantage women.
7. Academic organizational structures and rules contribute significantly to the underuse of women in academic science and engineering.
8. The consequences of *not* acting will be detrimental to the nation's competitiveness.[12]

The panel also addressed commonly held but false beliefs about women in science and engineering. In each case, they presented evidence refuting those beliefs. Table 2.1 compares beliefs versus the facts.

TABLE 2.1
Commonly held beliefs about women in science and engineering versus the facts

Belief	Evidence
Women are not as good in mathematics as men.	Female performance in high school mathematics now matches that of males.
The matter of "under-representation" on faculties is only a matter of time; it is a function of how many women are qualified to enter these positions.	Women's representation decreases with each step up the tenure-track and academic leadership hierarchy, even in fields that have had a large proportion of women doctorates for 30 years.
Academe is a meritocracy.	Although scientists like to believe that they "choose the best" based on objective criteria, decisions are influenced by other factors—including biases about race, sex, geographic location of a university, and age—that have nothing to do with the quality of the person or work being evaluated.
Women faculty are less productive than men.	The publication productivity of women science and engineering faculty has increased over the last 30 years and is now comparable to men's.

Source: Excerpted from Committee on Maximizing the Potential of Women in Academic Science and Engineering, National Academy of Sciences, National Academy of Engineering, and Institute of Medicine, *Beyond Bias and Barriers: Fulfilling the Potential of Women in Academic Science and Engineering* (Washington, DC: National Academies Press, 2007), table S-1, pp. 5–6.

Mathematics plays a key role in career development for both men and women. The best research on mathematics and career development is longitudinal. Longitudinal studies follow students over years, sometimes even decades. Clifford Adelman has reported data about the experiences of adults in their 30s who had been studied systematically since they attended high school.[13] The original sample was a nationally representative group of American high school seniors. In an intriguing analysis, Adelman compared the male/female salary gap for those people in their 30s who had studied mathematics in college as compared with those who had not. Among managers in financial institutions, women earned 29.1 percent less than men. However, when financial managers who had earned more than eight credits in college mathematics were compared, the women actually earned 4.5 percent more than the men.

As Adelman concluded, "More math meant more money."[14]

Pioneering Women Scientists and Mathematicians

The earliest women mathematicians had to overcome prejudice and insufficient recognition of their work. Both Sophia Germaine, who received the grand prize of the French Academy of Sciences in 1816 for her paper "Memoir on the Vibrations of Elastic Plates," and Mary Fairfax Somerville, who presented a paper entitled "The Magnetic Properties of Violet Rays of the Solar Spectrum" to the Royal Society of London in 1826 and was subsequently elected an honorary member of the Royal Astronomical Society, were self-taught.[15] Both hid their studies of mathematics from their families, who did not approve of their daughters studying such material. They received little or no formal training in mathematics. Somerville's study of mathematics was delayed during the three years she was married to a man opposed to the education of women. Many years later, the Women's College at Oxford was renamed Somerville College in recognition of her achievements. Because intelligent women were suspect, some women scientists used male or gender-neutral noms de plume. Sophia Germaine signed the name LeBlanc to her writings and letters. Augusta Ada Byron Lovelace, arguably the world's first computer programmer, published under the initials A.A.L.

Most students of statistics are not aware that one of the pioneers in this field was the nurse Florence Nightingale. Karl Pearson, in a letter to Sir Francis Galton, called her "the prophetess." Marti Rice and William Stallings have documented her accomplishments in the field of statistics, including calculating mortality rates, developing statistical records for hospitals, and conducting impact and evaluation studies.[16]

The philosopher Elisabeth of Bohemia (known as one of the "female Cartesians") used Cartesian arguments to challenge Descartes. Voltaire's lover, Emilie du Chatelet, questioned both Descartes and Isaac Newton. After translating *Principia* into French, she translated three more versions—"a simple prose account for newcomers, a presentation in the framework of the new continental algebra, and a summary of recent mathematical research and experimental vindications of Newton's theories (which demonstrated her own deep understanding of mathematics)."[17]

Despite these early achievements by women, students seldom learn of them. Stephen Brush has observed that

> the scientists and engineers mentioned or pictured in textbooks are almost always male. When a discovery made by a women is discussed, she is often not given credit. At the high school level, Marie Curie may be the only woman mentioned, perpetuating the belief that science has been created almost entirely by men. A striking example of the clumsiness of attempts to include women scientists is provided by a widely used high school physics text. C. S. Wu is pictured in the front of the book; the caption says she is well known for an experiment that disproved a basic principle of physics, but does not say what the experiment was. The experiment, which demonstrates that parity is not conserved in certain interactions of elementary particles, is discussed later in the same text, but Wu is not mentioned there![18]

In her classic book *Overcoming Math Anxiety*, Sheila Tobias reports on one of the earliest studies of the links between gender and mathematics:

> In 1974, John Ernest, a professor of mathematics at the University of California at Santa Barbara, put mathematics and sex on the public's agenda for the first time. Assigned to teach a freshman seminar about elementary statistics, Ernest decided to turn his seminar into an investigation of the relationships, real and imagined, between gender and performance in mathematics. His students fanned out into neighboring junior and senior high schools to interview teachers and students about girls' and boys' performance in mathematics. The results of their inquiry were nearly always the same. Both boys and girls, they were told, have a fair amount of trouble doing math, and most of them do not like the subject very much. The difference between them was that boys stuck with math, because they felt their careers depended on it and because they had more confidence than girls in their ability to learn it. The problem, concluded Ernest, was not math inability in females; it was their

math avoidance at crucial stages of their schooling. Society expects males to be better than females at mathematics. This affects attitudes; attitudes affect performance; performance affects willingness to study more mathematics; and, eventually, males do better than females. His findings, modest though they were, were considered so important that the Ford Foundation published and distributed forty thousand copies of his little book.[19]

Tobias has led the way in articulating, developing, and describing the concepts "math anxiety" and "math avoidance." She suggests that there are several internal, though not innate, variables that cause girls to do more poorly than boys in many mathematics courses and tests. One of these is "female isolation": "Unless they are blessed with a math-oriented family, or collected in a special dormitory for math/science majors, girls find themselves isolated both in class and outside of class when it comes to math. Not having anyone to talk to about what they're learning, they fail to learn to *speak mathematics*; worse yet, they do not get the opportunity to extend their knowledge, their skills, and their imagination through discussion."[20]

Building upon the work of Sheila Tobias, Cheryl Ooten and Kathy Moore have written *Managing the Mean Math Blues*, a book that presents practical skills, strategies, and exercises to help students confront and overcome math anxiety. They quote Albert Einstein: "It's not that I'm so smart, it's just that I stay with problems longer."[21]

One of Stephen Brush's final observations should be underscored: "When more women become successful scientists, the culture of science may change; it is very unlikely to change if women stay out of science."[22] Science progresses, as Thomas Kuhn has observed, when creative people shed the constraints of current theory to propose innovative shifts in our frame of reference—that is, our paradigms. One liability of a system that recruits only white males with a particular personality profile is that it restricts the range of potential creative solutions to scientific problems.

Women's colleges may provide a more encouraging and receptive environment for young women considering a STEM major. Etta Falconer, director of science programs at Spelman College, said about Spelman students, "We expect them to succeed and they do."[23] Success should be expected of both men and women, majority and minority students, across the United States.

False Assumptions about Students of Color

The achievement gaps in science and mathematics in the United States between under-represented minority students and white students have narrowed in the period since 1995, but they remain unacceptably large. In the 2003 National Assessment of Educational Progress (NAEP), the average science scores of African American fourth graders were 78 points below whites; Hispanics were 67 points below whites. For eighth graders, the gap was 89 points between African Americans and whites and 70 points between Hispanics and whites.[24]

The achievement gap in mathematics between ethnicities, although disturbing, is not as large. In 2003, the average NAEP mathematics scores of African American fourth graders were 27 points below whites and 6 points below Hispanics. For eighth graders, the gap was 36 points between African Americans and whites, and 7 points between African Americans and Hispanics. Table 2.2 presents the mean mathematics scores of fourth and eighth graders in 2003 and 2005.

The ethnic categories listed in Table 2.2 are the designations the U.S. government uses. But these categories exhibit flaws that plague much of evaluation research. Because a growing number of students identify with two or more ethnicities, it is a false construction to force a student to choose one ethnic category. In addition, categories like Hispanic, Latino, and Asian encompass many different specific cultures and nationalities. It may be more appropriate to place Pacific Islanders in the same category as American Indian, Alaskan Natives, and other indigenous groups as opposed to combining them with Asian and Asian American students.

In many fields of human endeavor, false perceptions about who has aptitude and who lacks it have created formidable barriers to access for talented people.

TABLE 2.2
Student scale scores in mathematics by ethnicity

	4th grade		8th grade	
Ethnicity	2003	2005	2003	2005
White, non-Hispanic	243	246	288	289
Black, non-Hispanic	216	220	252	255
Hispanic	222	226	259	262
Asian / Pacific Islander	246	251	291	295
American Indian / Alaskan Native	223	226	263	264

Source: National Assessment of Educational Progress. *The Nation's Report Card: Mathematics 2005*, National Center for Education Statistics (Washington, DC: U.S. Department of Education, Institute of Education Sciences, 2006).

Often these perceptions about aptitude are based on prejudices that certain ethnic groups or women or white people from poverty aren't smart enough.

Equity Lessons from the World of Sports

In the 20th century the color barrier was broken in many fields, including sports. Minority sports pioneers include baseball player Jackie Robinson and golfer Tiger Woods. One of the last barriers to fall was the exclusion of African Americans from playing quarterback in professional football. The first step was taken by the aptly named Willie "the Pro" Thrower, an African American who threw eight passes, completing three, for the Chicago Bears in October of 1953. In the 1970s, James Harris and Joe Gilliam were successful African American quarterbacks for the Rams and the Steelers, respectively.

In Super Bowl XXII, Doug Williams "exploded the quarterback myth for all time when he threw a record four touchdowns and guided the Washington Redskins to a 42-10 victory."[25] Williams threw all four touchdown passes in the second quarter. His total passing yardage in that game was 340 yards.

Even after Williams' stunning accomplishment, some believed that African Americans were not qualified to play quarterback, a position that demands intelligence and analytical thinking. In a brief stint as a commentator on *Monday Night Football*, Rush Limbaugh said about Donovan McNabb: "Sorry to say this, I don't think he's been that good from the get-go. I think what we've had here is a little social concern in the NFL. The media has been very desirous that a black quarterback do well. There is a little hope invested in McNabb, and he got a lot of credit for the performance of this team that he didn't deserve. The defense carried this team."[26]

We are now accustomed to black quarterbacks and superstars on professional sports teams. As Steve Chapman notes, "These days, in the eyes of the people who do the hiring and the firing in the NFL, there are no white quarterbacks or black quarterbacks. There are just good ones and bad ones."[27] But the prejudice barrier took time to break, and as in education, the barrier consisted of false allegations that African Americans were not as capable as whites.

Minorities and Admission Tests

Roy O. Freedle, a retired researcher formerly employed by the Educational Testing Service, demonstrated that the Scholastic Achievement Test (SAT) is biased both culturally and statistically against African Americans, Latinos, and Asian Americans.[28] His analysis of SAT data from the 1980s to the present reveals that the achievement gap between whites and these groups is greatly diminished if you consider only the difficult items on the SAT. *In other words, these ethnic groups do more poorly generally on the SAT because they do more poorly on the easy items, items that are more directly biased because they require cultural knowledge that may not be as readily available to minority students.* When you compare performance only on the really tough items, the achievement gap shrinks considerably or vanishes. When only difficult items are considered, the verbal score for some minority test takers increases by as much as 200–300 points.

On the basis of his findings, Freedle went on to develop a revised SAT (also known as the R-SAT) which essentially eliminated the 45 easiest questions.[29] He noted that

> fifteen high-frequency analogy words (such as "horse" and "snake") had an average of 5.2 dictionary entries, whereas rare analogy words (such as "vehemence" and "anathema") had an average of only 2.0 dictionary entries. Various researchers have hypothesized that each cultural group assigns its own meanings to such common words to encapsulate everyday experience in its respective cultures. Thus, individuals from various cultures may well differ in their definitions of common words. Communities that purportedly are speaking the same language may use the same words to mean different things.[30]

Freedle reported work by Fagan and Holland to develop a test free of racial bias. They selected vocabulary words not known by most people and used those unfamiliar words in a sentence so that an astute reader could gain a sense of the meaning of the word. Then they put the word into a multiple-choice test: "Under the assumption that racial differences in intelligence do *not* exist, Fagan and Holland reasoned that *if all individuals are given an equal opportunity to learn all the crucial concepts that are needed in order to select the correct answer* on any standardized test, there then should be *no* significant difference in how the races respond to the test. . . . The results confirm their hypothesis."[31] In fact, they found no racial differences on the test they constructed to be free of unintended racial bias. When the same study participants took a conventional IQ test (with its built-in bias), Fagan and Holland found racial differences.

In summary, African American students do more poorly on the easy questions from the SAT, questions that use seemingly common vocabulary words that might have a different meaning in the African American community than they do in the majority middle-class white community. The surprising finding from Freedle's research is that when the playing field is leveled (and when students with similar total scores are compared), neither white nor African American students seem to know esoteric words, and the African American students consistently do somewhat better than the white students.

Some minority students have trouble learning math and science because scientists seem to have their own language. To complicate matters, the language in one scientific specialty differs from that in another specialty. Bryan A. Brown of Stanford University notes that "science has developed a way to convert complex processes that are generally described through clauses into simple nouns that represent complex ideas in a concise manner." He cites as examples processes like photosynthesis and glycolysis: "Once this rhetorical reconstruction has been accomplished, this style of scientific communication provides its users with a venue where complex ideas can be readily exchanged if users *share the same interpretive framework*. . . . This practice . . . creates a discursive environment in which the language of science has evolved into a lexically dense, impersonal, fact-driven type of discourse."[32] Brown also notes that the language of science is relatively unfamiliar to African Americans and other underrepresented minority students and that this language discontinuity is a partial explanation for the achievement gap.

Daryl Smith and Gwen Garrison conducted an outcomes-based review of the impact of standardized admission tests, specifically the SAT, the GRE (Graduate Record Examination, the standard evaluation for admission to graduate school), and the LSAT (the Law School Admissions Test).[33] Noting that most predictive studies about these tests focus on the first year of college, graduate school, or law school, these two authors chose to look at long-term, more meaningful outcomes, such as graduation rates. Table 2.3, extracted from their report, presents the graduation rates associated with SAT score categories.

Smith and Garrison concluded that had colleges eliminated students within the lowest category of SAT scores—those with scores of less than 900—they

TABLE 2.3
Selective university system-wide data: Combined SAT scores and graduation rates

Combined SAT score	Graduation percentage
<900	64
900–1100	71
1100–1300	78
1300–1600	82
All SAT scores	75

Source: Excerpted from Daryl G. Smith and Gwen Garrison, "The Impending Loss of Talent: An Exploratory Study Challenging Assumptions about Testing and Merit," *Teachers College Record* 107, no. 4 (2005).

STEREOTYPE THREAT

Social psychologist Claude Steele conducted a series of rigorous experiments to investigate a phenomenon he calls "stereotype threat," which he believes fully explains the gap on IQ test achievement between whites and African Americans (and, for that matter, between young men and women). He argues that African American students are well aware of the stereotype and expectation that they are not as intelligent as whites and that they will perform poorly. This awareness generates anxiety that depresses their performance on such tests. His experiments are elegant, rigorous, and persuasive.*

In one study, Steele and colleague Joshua Aronson compared a sample of white college students with a comparable sample of African American students. Both were given a series of tough questions from the GRE, and the students were told the source of these questions. The usual achievement gap appeared in the results. Steele then gave the same test battery to two other groups of white and black college students. This time, however, he introduced the test items as questions that might be used in a future test. When the pressure was lifted because of this introduction, no achievement gap between the two groups appeared.

In a second study, Steele found an achievement gap between white and African American college students under one condition, but not under a second condition. First, the students filled out a background sheet like those usually associated with the SAT or the GRE exam, which included a question on ethnic background. Under the second condition, there were no background questions. Merely asking a minority student to identify his or her ethnic background generates stereotype threat and suppresses the student's score. (Steele controlled for SAT score in each of these studies.)

*Both of the studies described here are reported in C. M. Steele and J. Aronson, "Stereotype Threat and the Intellectual Test Performance of African Americans," *Journal of Personality and Social Psychology* 69 (1995): 797–811.

"would have eliminated 64 percent who are successful [in graduating from college] in order to avoid admitting 36 percent who are not successful. Even in the highest SAT group, nearly 20 percent were not successful. How to evaluate this level of risk would depend upon an institution's own view of the students in this lowest category and the other attributes these students might bring to the institution. For these data, scores below 900 could be considered a threshold."[34]

Based upon a similar analysis of the value of the LSAT in predicting passage of the bar exam, Smith and Garrison found that "eliminating the 13 percent of the population with the lowest test scores would have eliminated 75 percent of students from those groups who succeeded."[35] In other words, even the students who scored lowest on the LSAT did approximately as well as the rest of the students when it came to the ultimate test—passing the bar exam.

In a similar vein, the American Association of University Women reports that "the Massachusetts Institute of Technology found that a woman with the same SAT score as a man was likely to get better grades. After adjusting its admissions process to compensate for the SAT's 'under-prediction,' MIT has found that its women students earn higher GPAs in more than half of majors, even though their average SAT-math score is 20–25 points lower than that of their male peers."[36]

Judgments about Intelligence

> There is an old tale of two friends who meet on the street after not seeing each other for 25 years. One, who had graduated at the top of his class, was now working as an assistant branch manager of the local bank. The other, who had never overwhelmed anyone with his intellect, owned his own company and was now a millionaire several times over. When his banking friends asked him the secret of his success, he said it was really quite simple. "I have this one product that I buy for two dollars and sell for five dollars," he said. "It's amazing how much money you can make on a 60 percent markup."
>
> Mark McCormack, *What They Don't Teach You at Harvard Business School*[37]

Many of us were educated to believe that intelligence is a fixed character trait like height, weight, and hair color. Intelligence can be expressed as an IQ number; its value, generally between 80 and 120, is largely an inherited characteristic. But recent sociological and psychological research about intelligence and aptitude, including my own research, has suggested that this concept about innate intelligence is wrong. Yes, some people are smarter than others. But the latest studies indicate that there are a number of different kinds of intelligence and a person may be high on one kind and low on another. This is one of the reasons that some people who did poorly in high school go on to be great successes in business.

The idea that there is one fundamental trait called intelligence is rapidly being consigned to the landfill of intellectual history. Stephen J. Gould largely disproved the unitary intelligence concept derived from psychological research conducted by Charles Spearman in the first decade of the 20th century. Gould traces how Spearman's conclusion that there was one underlying general intelligence trait (which Spearman modestly called "Spearman's g") was an artifact of his application of a statistical technique (principal components factor analysis) that was mathematically predetermined to generate one large factor or intelligence concept. Other statistical techniques (such as other forms of factor analysis based on different assumptions) would have yielded two or three "medium-sized factors."[38]

Historically, elites have justified their positions of power by asserting that the less powerful lack the intelligence to succeed in society. The early Roman conquerors believed that the inhabitants of what is now Great Britain were much less intelligent than Romans. Nineteenth-century European imperialists mocked the alleged intellectual inferiority of colonial populations. As I discuss below, a prominent British scientist argued that empirical evidence from tests of aptitude revealed that Jewish people were not intelligent enough to benefit from schooling. At one time, it was popular in scientific circles to measure the brain size of people from various ethnic groups.

Assumptions about intelligence and who possesses it often are at the heart of debates about education. Such assumptions include the notion that highly intelligent men marry highly intelligent women and are likely to succeed economically, and that poor children are not as intelligent and more likely to fail in school. Why invest so heavily in educational resources for African American children when they are not as intelligent as majority children? Why not just accept the gender gap in mathematics and acknowledge that girls cannot do as well as boys?[39]

Such unjustified assertions are at the core of the negative expectations about mathematics and science education that permeate our society. They go a long way toward explaining why some groups are underrepresented in the study of mathematics and science (as well as other fields). In this book I would like to isolate, analyze, and refute the assumptions implicit in such assertions. Consider:

- *Even if these assumptions were true,* they would not change the fundamental arguments and recommendations in this book.
- Fewer and fewer social scientists support the idea of a unitary "intelligence." Attempts to measure head and brain sizes were rejected by most scientists years ago.
- Intelligence is not the main reason people succeed.
- Discussion of differences in the intelligence levels of population subgroups usually involves the notion of inherited aptitude. Arguments based

on this notion display considerable confusion about the manner in which genetics and environment interact.

What If the Assumptions Were True?

What if science had proven that one gender or ethnic group was superior to another in native intelligence? I would still advocate that mathematics and science education be provided for all students. Even if differences had been shown to exist, the variation between groups would be small relative to the variation within groups. Consequently, if such assertions were true—say, in the case of males and females—there would be two overlapping normal curves, one for young men and one for young women. A tiny percentage of those in the normal male curve would have a higher IQ than that of the highest woman, and a tiny percentage of those from the female normal curve would have an IQ below that of the lowest scoring male. The vast majority of both young men and young women would fall in the same IQ range.

Aptitude is but one of many factors that affect whether a student masters a subject and applies that knowledge in subsequent studies and in the workplace. Learning is affected by the health of the student, the quality of the teaching, the nature of the curriculum, and other factors discussed at length in this book. Expectations and self-concept play a vital role. *So does hard work.*

While there is no convincing evidence that one group is more intelligent than another, *even if there were,* my conclusions and recommendations about mathematics and science education would remain unchanged.

Rejection of a Unitary Concept of Intelligence

Current-day psychological and educational researchers like Howard Gardner, whose work is considered to be on the cutting edge, have concluded that there are multiple dimensions of intelligence.[40]

Because the concept of intelligence is not epistemologically sound (in fact, it may be a nominalistic fallacy), there sometimes are strange research results. James Flynn was the first to observe and report that IQ scores around the world have been steadily increasing over time.[41] This phenomenon has now been dubbed the Flynn effect. Cognitive psychologists, researchers, and policymakers aren't sure why this is happening. In a recent work, Flynn suggests that IQ tests measure cognitive skills that can be taught and learned and, furthermore, many societies have become more adept over time at socializing their children with these skills.[42] In short, there is increasing evidence that IQ is not an innate genetic trait.

Intelligence Is Not the Main Cause of Success

In his bestseller *Outliers,* Malcolm Gladwell explores the factors that contribute to success.[43] He concludes that talent plays a minor role. He provides some interesting examples, including the leading amateur and professional hockey players in Canada. Virtually all of them were born in the first three months of the year. He links this intriguing fact (which seems to imply that babies born in January, February, and March are better hockey players) to public school rules about which grade a child is placed based on his birthday. Children born early in the year become the oldest children in their class. Therefore, they tend to be larger and stronger than their classmates. They are recruited more frequently as rising young future stars in hockey and *are given the best coaching, support, and guidance.* So hockey success in Canada is less a function of innate talent and more attributable to other factors, such as birth date.

Gladwell also cites examples of successful people who devoted long hours to learning and perfecting their craft. He notes that Bill Gates devoted day and night to working and playing with computers while a high school student in Seattle. He attributes the success of the Beatles to their two-year stint in Hamburg, Germany, a period during which they were onstage in strip clubs playing their music eight hours a day, seven days a week. Gladwell concludes that extreme success in any field requires at least 10,000 hours of practice.

David W. Galenson is an economist who began visiting art galleries as a PhD student to take a break from cramming for his comprehensive examinations and, later, from working on his dissertation. Having continued to visit museums and galleries long after receiving his degree, he became interested in the ages at which famous artists produced their greatest works, which he defined as the works selling for the highest prices. Galenson then produced a series of scatterplots, or two-dimensional charts. The *x*-axis was the age of the painter. The *y*-axis was the price of a painting produced at that age. When he plotted Picasso's paintings, the graph was high when Picasso was a youth and declined with age. When he plotted Cezanne's paintings, the graph started low and slowly increased with age.[44]

Galenson found the same two patterns repeatedly when he looked at other artists and when he looked at creative people in other fields—for example, movie directors. He believes he has uncovered two enduring patterns of creativity. Those who follow the Picasso pattern are brilliant young people who produce their greatest efforts early in life. They continue to work and to create, but to some degree, they are riding on the laurels of their early success. Those who follow the Cezanne pattern gradually accumulate knowledge and experience and continuously incorporate what they have learned into their work. As a result, their work becomes better and better as they age.

I believe that most people—and to bring us back to the focus of this book, most STEM students—follow the Cezanne pattern.

The harder you work, the more homework you do, the better you will perform in math and science.

Arguments like Gould's about the epistemological implications of analytical assumptions constitute a special case of arguments about the links between mathematics and reality. Some would argue that mathematical elegance and beauty underlie the seemingly chaotic physical reality we observe and would attribute theological characteristics to mathematics. Galileo once remarked that mathematics is the language with which God wrote the universe. I know a statistics professor who asks each of his students to gather a thousand maple leaves and measure the distance between the first and third points of the leaf. He then plots and graphs the combined data points from all students in the class to demonstrate that the result is a perfectly defined statistical normal curve. Some students are awestruck when they first discover the precise mathematical relationships underlying physical phenomena.

The perpendicular coordinates utilized in these analyses are referred to as "Cartesian coordinates," after the French mathematician René Descartes. The French reify mathematics. Several years ago I attended an international conference at a French technical institute in a suburb of Paris. The purpose of the meeting was to assess the relative merits of objective, multiple-choice tests and the traditional French baccalaureate examination in college admissions. At one point I engaged in a debate with the dean of one of France's leading medical schools. He stated flatly that since his school wanted to admit only the most promising students, rigorously selected, there was only one criterion for admission, namely, the student's score on a demanding mathematics test! Much as I like mathematics, I replied that if I were undergoing surgery, I would hope that the surgeon had demonstrated skills and knowledge in other areas besides mathematics when he or she entered medical school.

Intelligence Is Not Related to Race or Ethnicity

The problems and solutions described in this book apply to middle-class white males just as much as to minority groups. Still, poor people, women, and people of color each face additional barriers. Furthermore, the obstacles accumulate rapidly for a person who falls into two or three of these categories—for example, a poor African American woman. Ultimately, all of the students who are "underrepresented" in the study of mathematics and science represent an enormous potential resource for the U.S. economy.

When the barriers to studying and achieving in mathematics and science are destroyed, the technical skills and knowledge available in the American workplace will increase dramatically. The removal of these barriers can transform lives.

The barriers disadvantaged groups face are defended and justified by a false assertion: "Let's face it—they are not as intelligent as middle-class white males."

The eminent statistician Karl Pearson, who had analyzed data from intelligence testing of children in the first decades of the 20th century, discussed at length how Jewish people should be barred from immigrating to the United Kingdom because of their inferior intelligence. The subsequent intellectual contributions of Jewish scholars—as reflected, for example, in the number of Jewish Nobel prize winners—provide an easy refutation of Pearson's thesis. But his thesis was taken seriously at one time—he was, after all, a leading scholar—and long-term damage had been done. According to Elazar Barkan, Pearson's study concluded that

> Jewish girls were distinctly less intelligent than Gentile girls, whereas Jewish boys scored worse than Gentiles in good schools but better than those in poor schools. Faced with statistics which could be variously interpreted, Pearson quickly concluded: "Our alien Jewish boys do not form from the standpoint of intelligence a group markedly superior to the natives. But that is the sole condition under which we are prepared to admit that immigration should be allowed." Aggregate average placed the Jewish population rather low physically and mentally, besides which no one could say how adaptable their progeny would become. Despite the risk that Britain may "exclude a future Spinoza," any further immigration would violate "the law of patriotism."[45]

In the 1990s the book *The Bell Curve: Intelligence and Class Structure in American Life,* by Richard Herrnstein and Charles Murray, argued that African Americans are less intelligent than whites.[46] Their argument received a considerable amount of attention. I believe Herrnstein and Murray drew the wrong conclusions from their data. But worse than that, they asked the wrong question.

Instead of asking what differences might be used to exclude some groups of people from education, wealth, and status, we should be asking how we can educate all youngsters so that they can compete in an increasingly technological economy.

Unfortunately, Herrnstein and Murray's thesis that African Americans are intellectually inferior to whites became accepted by many readers and lay people—even though the thesis is completely erroneous. The conclusions of *The Bell Curve* reinforced stereotypes and prejudices held by many people and persuaded others that their own egalitarian views were scientifically unjustified. It became acceptable to speak about the intellectual inferiority of African Americans.

After the publication of the book, discovering errors and flaws in Herrnstein and Murray's work became a growth industry among academics and policymakers. It is important to understand why their argument is wrong and to respond on scientific, not emotional, grounds. Their work contains statistical flaws, in addition to complete reliance on a unitary measure of intelligence and their failure to consider stereotype threat. (The concept of stereotype threat by itself could explain the black/white differences in IQ scores in *The Bell Curve*.)

Early chapters in *The Bell Curve* sought to establish the importance of intelligence with respect to important adult outcomes. The authors present many tables reporting the results of logistic regression analyses. Dependent variables include education level and income. Among the independent variables, intelligence inevitably emerges as a strong predictor. Few reviewers or readers have studied multiple regression. Only some of those have studied logistic regression. To some, it seems like a mathematical advance over ordinary multiple regression. The key distinction between the two kinds of regression is buried deep in an appendix to this voluminous book: "All dependent variables are binary."[47] In short, when these authors predicted income, they were not really predicting the full interval variable income. Rather, they created a dichotomous, or binary, dependent variable indicating whether or not each person was above or below the poverty line. Similarly, when Herrnstein and Murray claimed they were predicting education, they really were predicting whether or not each person had a college degree. Their use of dichotomous dependent variables and, therefore, their use of logistic regression, greatly weakened their analysis.

Furthermore, Herrnstein and Murray made no attempt to consider or control for multicollinearity. This is a condition present in some correlation matrixes which can degrade, and sometimes render meaningless, multiple regression analyses. It occurs when two or more of the predictor variables are very highly correlated. Herrnstein and Murray consistently worked with predictor variables that conceptually might be expected to exhibit multicollinearity. However, they failed to acknowledge, discuss, confront, or control for this potentially devastating mathematical problem in their data.

Hidden Talent in Impoverished Groups

Throughout history, dominant cultures have attempted to justify their subjugation of other groups by asserting that those groups were mentally inferior. Take the European colonial powers' subjugation of people of color, for example. The emergence of brilliant thinkers from minority populations has forced a reevaluation of the thesis that cultural dominance is preordained according to race.

For many decades, British intellectuals deprecated the intellectual capabilities of people from India; given this negative stereotype, the contributions of east Indians to the subjects of mathematics and statistics are even more impressive. For example, the statistician P. C. Mahalanobis made fundamental contributions to the development of multiple group discriminant function analysis, including defining a complex multivariate generalized distance measure (Mahalanobis's D^2). But few stories in the history of science and mathematics can match that of the short, troubled life and extraordinary accomplishments of Srinivasa Ramanujan. Graduate students in mathematics today carefully study Ramanujan's theorems. Although he had a brief professional career and died young, Ramanujan is recognized as a towering major figure in the history of mathematics.

Ramanujan's life illustrates how talent can be hidden in unlikely places. A sickly youth, Ramanujan grew up in an ordinary village in southern India. He stumbled through school, flunked out of college, could not keep a job, and had a troubled marriage. But he discovered mathematics as an adolescent, devouring a British volume he acquired, Loney's 1893 textbook *Trigonometry*. He became obsessed with mathematical games, puzzles, and theorems and recorded them in notebooks he carried around South India as he lurched from failure to failure.

In 1913 he asked questions of three of the world's leading mathematicians at Cambridge University in England. One, G. H. Hardy, immediately recognized the genius that had produced these questions. A correspondence and then a friendship ensued. Hardy eventually arranged for Ramanujan to travel to Cambridge, where he impressed the resident math scholars. However, his poor health continued, compounded by isolation from his family in India and his inability to adjust to British life. He died at the age of 32. In his powerful biography of Ramanujan, *The Man Who Knew Infinity*, Robert Kanigel writes:

> "Srinivasa Ramanujan," an Englishman would later say of him, "was a mathematician so great that his name transcends jealousies, the one superlatively great mathematician whom India has produced in the last thousand years." His leaps of intuition confound mathematicians even today, seven decades after his death. His papers are still plumbed for their secrets. His theorems are being applied in areas—polymer chemistry, computers, even (it has recently been suggested) cancer—scarcely imaginable during his lifetime. And always the nagging question: What might have been, had he been discovered a few years earlier, or lived a few years longer?[48]

Kanigel notes how tenuous and serendipitous the events that led to Ramanujan's recognition were and speculated that things might have turned out differently:

> It is a story of one man and his stubborn faith in his own abilities. But it is not a story that concludes, Genius will out—though Ramanujan's, in the main, did.
>
> Because so nearly did events turn out otherwise that we need no imagination to see how the least bit less persistence, or the least bit less luck, might have consigned him to obscurity. *In a way, then, this is also a story about social and educational systems, and about how they matter, and how they can sometimes nurture talent and sometimes crush it.*[49]

Twenty years after Ramanujan's death, E. H. Neville observed:

> Ramanujan's career, just because he was a mathematician, is of unique importance in the development of relations between India and England. India has produced great scientists, but Bose and Raman were educated outside India, and no one can say how much of their inspiration was derived from the great laboratories in which their formative years were spent and from the famous men who taught them. India has produced great poets and philosophers, but there is a subtle tinge of patronage in all commendation of alien literature. Only in mathematics are the standards unassailable, and therefore of all Indians, Ramanujan was the first whom the English knew to be innately the equal of their greatest men. The mortal blow to the assumption, so prevalent in the western world, that white is intrinsically superior to black, the offensive assumption that has survived countless humanitarian arguments and political appeals and poisoned countless approaches to collaboration between England and India, was struck by the hand of Srinivasa Ramanujan.[50]

The same idea was expressed in 1946 by the Indian statesman Jawaharlal Nehru: "Ramanujan's brief life and death are symbolic of conditions in India. Of our millions how few get any education at all; how many live on the verge of starvation. . . . If life opened its gates to them and offered them food and healthy conditions of living and education and opportunities of growth, how many among these millions would be eminent scientists, educationists, technicians, industrialists, writers, and artists, helping to build a new India and a new world?"[51]

Late in life, Hardy wrote: "The real crisis of my career came . . . in 1911, when I began my long collaboration with Littlewood, and in 1913, when I discovered Ramanujan. All my best work since then has been bound up with theirs, and it is obvious that my association with them was the decisive event of my life."[52]

Srinivasa Ramanujan's incredible journey and brilliant contributions to mathematics remind us that talent, even genius, can be found in the most unlikely corners of society. We must find, nourish, and unleash all this hidden talent, which can transform our society and our economy.

Robert Sternberg, provost of Tufts University, is considered to be one of the leading U.S. authorities on intelligence and intelligence testing. He has published many books and articles on this subject. In an article in the *American Scholar,* he revealed why the subject fascinates him.

> The battle is a personal one. As a child, I did horribly on IQ tests. As a result, my teachers in the early years had low expectations, and I gave them what they wanted. They were happy and so was I. By fourth grade, I had a teacher who expected more, and I gave her more. I went from being a B–C student to being an A student. I attributed what success I came to have to going to bed early. I knew my IQ was low—what else could it be? Now, as a theorist and a researcher in the field of intelligence, I would like to think that there may have been more involved. More of what? Well, that's what I'm still trying to find out.[53]

CHAPTER 3

Effective Leadership, Careful Evaluation

It is early morning. A woman stands in the doorway of the bedroom and says, "It's time to get up and go to school."

A figure huddles under the covers with a pillow pulled over his head. In a muffled voice he says, "I don't want to go to school."

The woman says, "You have to, and it's getting late."

He replies, "I hate school. None of the teachers like me. All the kids make fun of me."

In exasperation, she says, "But you've got to get up and go to school. You're the principal."

As discussed in chapter 1, in the current high-tech global economy STEM education is of vital importance. Yet international assessments of math and science achievement have consistently placed American students below their counterparts from most other nations. Furthermore, the science career ladder for women in the United States has hit a glass ceiling, and a persistent achievement gap exists between white and minority students.

That's the lemon. How do we turn it into lemonade?

Effective leaders create successful institutions. Building upon the insights of the economist and management guru Peter Drucker, I examine schools that have excelled—even though they would have every excuse to fail—and present some inspiring case studies of leadership in K–12 and higher education.

Reform requires strategies for change. But proposed innovations should never be implemented widely unless we know they work and know which components of reform work best. Successful reform requires systematic, hard-nosed evaluation. It is no longer acceptable to change policy and practice based merely on good intentions and anecdotal evidence. I present a blueprint for effective program evaluation of educational initiatives and discuss how poorly designed evaluation and assessment under the No Child Left Behind Act have compromised educational progress.

Knowledge Workers

STEM professionals are knowledge workers—they are paid for what they know rather than for what they do. "Knowledge worker" is a term that Peter Drucker coined in his 1959 book *Landmarks of Tomorrow.* Nearly four decades later, he said: "By the end of this century knowledge workers will make up a third or more of the work force in the United States—as large a proportion as manufacturing workers ever made up, except in wartime. The majority of them will be paid at least as well as, or better than, manufacturing workers ever were. And the new jobs offer much greater opportunities."[1]

Yet Drucker also saw the flip side of this change: "But—and this is a big but—the great majority of the new jobs require qualifications the industrial worker does not possess and is poorly equipped to acquire. They require a good deal of formal education and the ability to acquire and to apply theoretical and analytical knowledge. They require a different approach to work and a different mindset. Above all, they require a habit of continuous learning."[2]

Drucker began his book *The New Realities* by noting that "sometime between 1965 and 1973 we passed over . . . a divide and entered the next century. We passed out of creeds, commitments, and alignments that had shaped politics for a century or two."[3] Spotlighting an important trend, Drucker saw that "the shift to knowledge and education as the passport to good jobs and career opportunities means, above all, a shift from a society in which business was the main avenue of advancement to a society in which business is only one of the available opportunities and no longer a distinct one." Society's center of gravity, he observes, "is shifting to a new group—the knowledge worker—who has new values and expectations."[4]

In this book and elsewhere, Drucker chronicled the rise and fall of the blue-collar worker. Farming once dominated the American economy, but early in the 20th century the farming sector began a decline that never leveled off. Similarly, blue-collar factory workers provided the labor that fueled the global economy from the Industrial Revolution to the 21st century; however, factory work in the United States has begun to decline. Blue-collar assembly lines still create many of the products Americans buy, but increasingly, assembly work is moving to countries where labor costs are considerably lower. According to Drucker, "If any phenomenon can compare to the rise of the blue-collar, mass-production worker in this century, it's the fall of the blue-collar, mass-production worker."[5]

Drucker felt that education was the key to a better society, one with knowledge, not products, as its main economic focus. Schools, then, were essential in

ABOUT PETER DRUCKER

The obituaries at the time of Peter Drucker's death presented the astounding array of his professional accomplishments: creating and defining the field of management, framing it as a "liberal art," articulating the importance of non-profits and of management for nonprofits, providing critical advice to multinational corporations. Some of these articles quoted Jack Welch on how Drucker's key questions and advice had shaped the vision, goals, and strategies of Welch and of General Electric.

Drucker had deep knowledge of a wide range of subjects. Of course, he was the world's acknowledged expert on management. He also taught courses on the side about Japanese art at Pomona College. And he published two novels when he was in his 70s. Much of his writing began with his profound understanding of European history, especially the history of ideas.

Drucker discussed student intelligence at length. He questioned the meaning of international assessments that showed Japanese students outperforming U.S. students handily and discussed the Japanese "cram schools." He asked, "How would the scores compare if the tests were given to these same students ten years after they graduated from high school?"

Repeatedly he identified trends and forces in our society long before anyone else. For example:

A few years ago Malcolm Gladwell wrote a *New Yorker* article about the impact of dependency ratios and pension plans on national economies. He discussed, for example, how the economy of Ireland had grown powerful in recent decades because of a shift in dependency ratios caused by changes in birth control practices. He noted that the 21st-century impact of pension plans on our own economy had been predicted perfectly—in an article Drucker published in 1950! Gladwell said of this article, "The Mirage of Pensions," published in *Harper's Magazine,* that it

> ought to be reprinted for every steelworker, airline mechanic, and autoworker who is worried about his retirement. Drucker simply couldn't see how the pension plans on the table at companies like G.M. could ever work. "For such a plan to give real security, the financial strength of the company and its economic success must be reasonably secure for the next forty years," Drucker wrote. "But is there any one company or any one industry whose future can be predicted with certainty for even ten years ahead?" He concluded, "The recent pension plans thus offer no more security against the big bad wolf of old age than the little piggy's house of straw."*

*M. Gladwell, "The Risk Pool," *New Yorker,* Aug. 26, 2006.

It sometimes seemed that Drucker had known everybody of political, intellectual, or artistic importance, from Ludwig Wittengenstein to Martha Graham to Yogi Berra. He even met Sigmund Freud. Granted that, when they met, Freud was old and Drucker was a child. Freud had been an associate and a dinner guest of Drucker's parents.*

Drucker moved to England early in his career. The first book he wrote, *The End of Economic Man: The Origins of Totalitarianism,* became a best-seller in the late 1930s after it received a laudatory review in the distinguished *Times of London Literary Supplement.* The reviewer was Winston Churchill, who observed, "Mr. Drucker is one of those writers to whom almost anything can be forgiven because he not only has a mind of his own, but has the gift of starting other minds along a stimulating line of thought."† Later, as prime minister, Churchill ordered that this book be given to all officers graduating from a British military academy.

Consider these contrasts with most contemporary academics:

- Well into his 90s, he continued to write on a typewriter.
- He had no Federal or other research grants.
- He had no research assistants.
- He had no secretary.

*Personal Communication, circa 1985. I once mentioned the Nobel Prize in Economics to him. He replied: "A great award. Five hundred years too soon. But a great award."
†Gladwell, "Risk Pool."

the delivery of quality teaching and learning needed in the new knowledge economy: "The fourth-grade teacher whom I still remember once said many years later that . . . the job of the teacher is to find the strengths of the student and put them to work, rather than to look at the student as somebody whose deficiencies have to be repaired."[6]

Schools and colleges are complex organizations. Highly effective school principals and college presidents who apply leadership skills can generate success, even in contexts where you would never expect it.

Unexpected Success in Elementary and Secondary Education

The residents of a remote Italian village called Limone sul Garda consume a diet rich with pasta, butter, and saturated fat. Yet when doctors examined Valerio Dagnoli, a man who had dangerously low levels of HDL, the good cholesterol, they were surprised to find that he had no symptoms of heart disease. (HDL removes bad cholesterol and plaque from the coronary arteries. More HDL is better.) Many men in Limone sul Garda live long, healthy lives with very low rates of cardiovascular illness. Research revealed that the men of Limone sul Garda had a rare form of HDL, called ApoA-1 Milano. Scientists are now working to create a synthetic form of this extraordinary kind of HDL. One medication is currently undergoing extensive experimental field testing by the National Institutes of Health. An article about this promising pill called it "Drano for the heart." There is the real possibility that the chain of innovation that began with Valerio Dagnoli may eventually eradicate cardiovascular disease.[7]

AIDS researchers have conducted many studies in Africa, where this dreaded disease is epidemic. One group of scientists has focused on a small number of prostitutes in the slums of Nairobi. Shockingly, the sex often is unprotected and often is with men who are HIV-positive. As a result, many prostitutes develop AIDS. But surprisingly, some do not! Medical experts are totally puzzled by the women who are resistant to this deadly disease. Teams of researchers have flocked to Kenya to study them. Is it possible that these unlikely models of health could lead us to a cure for AIDS?[8]

Cures may be discovered simply by studying surprising examples of health where we expect to find illness. Similarly, we can learn about rethinking and repairing our educational reform by studying schools and students that have succeeded despite having every good reason to fail. Consider seven studies of schools that have achieved astonishing success. The same emphases, values, and strategies emerge in these success stories. And, in each case, much of the credit must go to school leadership, including leadership by both the principal and master teachers.

1. Reeves Study

Dr. Douglas Reeves found five characteristics in 90-90-90 schools (schools where 90% of students are poor, minorities, and yet still performing at grade level):

1. a focus on academic achievement
2. clear curriculum choices
3. frequent assessment of student progress and multiple opportunities for improvement
4. emphasis on writing
5. external scoring[9]

2. *Breakthrough High Schools*

The National Association of Secondary School Principals conducted the Breakthrough High Schools Project in 2002, focusing on schools that were at least 50 percent minority, at least 50 percent poor, and in which at least 90 percent of their students graduated and went on to postsecondary education. These breakthrough schools shared four characteristics:

1. a personalized environment that is physically and intellectually safe
2. a rigorous curriculum that engages students
3. collaboration among students, staff, parents, and community through a variety of extended learning opportunities
4. effective leadership—distributive leadership, shared decision making, and collaborative problem solving[10]

3. *Collaborative High-Performance Study*

In a two-year study, researchers from EdSource, Stanford University, the University of California–Berkeley, and the American Institutes for Research sought to find out why some schools performed better than others despite serving similar students.[11] They found that a set of practices were associated with high performance and grouped those practices into six categories:

1. Student achievement is a priority (i.e., expectations are high).
2. A coherent, standards-based curriculum and instructional program has been implemented.
3. Assessment data are used to improve student achievement and instruction (i.e., evaluation and decision making are data-driven).
4. The availability of instructional resources is ensured.
5. Principal leadership in the context of accountability-driven reform is being redefined to focus on effective management of the school improvement process.
6. District leadership, accountability, and support appear to influence student achievement.

Two people interviewed in this study made telling comments about how the high-performing schools and school systems work:

> I think the answer lies in the personnel: I mean the principal and the teachers doing what they need to do for children, using the data, implementing the

programs. I use the word "relentless" a lot. They've got to be relentless about not accepting anything but learning from the children. They're not going to let the children fail; they're going to make them learn. (7)

There isn't an elementary principal in this district who doesn't know what's expected of them, what their work plan needs to have, . . . and what's going to happen when my deputy superintendent and I do tours of that campus. They know what we're looking for. (15)

4. Urban Schools Study

Linda Darling-Hammond and Olivia Ifill-Lynch worried that "urban U.S. high schools are often factories for failure."[12] To find formulas for success, they studied schools in New York and Boston serving low-income, minority students but with college-going rates of more than 90 percent. They found that teachers in those schools worked together to create a collaborative learning environment. The teachers focused on the reasons students don't do their homework and then developed strategies to confront those barriers. Debra Meier, a senior scholar at New York University and founding principal of successful schools in New York and Boston, said: "We soon realized that a sizable number of students didn't really know how to do the homework, or at least how to do it well enough to get any satisfaction from it. A smaller number truly didn't have time, and we needed a whole-family conference to tackle the issues of jobs, baby-sitting, etc. A third group just couldn't or didn't plan, so we tried having a brief meeting at the end of each day to plan for homework. Some students were just expressing their general despair this way."[13]

Successful teaching strategies included creating homework assignments based on longer class projects and having students report on their homework in class the next day.

5. Chenoweth Study

Karin Chenoweth conducted a careful study of schools in neighborhoods characterized by both poverty and minority populations that excelled academically. She selected schools across the country that met the following criteria:

- A significant population of children living in poverty and/or a significant population of children of color.
- Either very high rates of achievement or a very rapid improvement trajectory.

- Relatively small gaps in student achievement in comparison with achievement gaps statewide.
- At least two years' worth of data.
- In the case of high schools, high graduation rates and higher-than-state average promoting power index (PPI).
- Adequate Yearly Progress (AYP).
- Open enrollment for neighborhood children; that is, no magnet schools, no exam schools, no charter schools.[14]

Chenoweth wanted to find out why these schools succeeded while neighboring schools failed. Certainly, part of the difference in achievement was related to a difference in attitude. At underperforming schools, she found attitudes like this:

> I had talked with teachers in crummy schools who told me, with great condescension, "These kids aren't like your children," meaning that most of the children they taught couldn't be expected to learn as much as white, middle-class children of college graduates. They would often add that advising their students to go to college was a waste of time. (2)

In contrast, in the high-achieving schools, she found that

> at these schools, just about all children meet or exceed state standards or are rapidly moving toward that goal. At these schools many, if not most, of the children are poor, and many, if not most, are children of color. Some of these schools are in neighborhoods that many middle-class parents would never consider allowing their children to set foot in. Some would say these schools could never be expected to teach their students to high standards. And yet the teachers and principals in these schools are demonstrating that, by carefully organizing their time and resources, they can make sure that their students learn to "read, write, and cipher," as on old fashioned educator said to me—and much, much more. (3)

Principals and teachers in nearby schools where the students were not achieving often developed elaborate explanations as to why the schools Chenoweth studied were excelling: "One principal told me with great amusement that other principals in her district had decided that her town had always had a 'strain of brilliant children,' as though academic achievement were something of a genetic disease" (14). In contrast, Chenoweth found that while the teachers in the achieving schools prepared students for statewide tests, they did not teach to the test.

> Although teachers spend a bit of time exposing students to test formats and to the way test questions are formulated, they say they do not concentrate on "test prep." "I was talking to a colleague in another district, and he said, 'So you teach to the test,'" said assistant principal Patrick Swift. "I said no, to meet state standards." "It's not just for testing" is how the science department chair, Recchia, put it. "You want them to know how to interpret, compare, analyze—not for the test but to be productive members of society." (185)

In Port Chester, New York, principal Carmen Macchia is seen as a key to the success of Port Chester Middle School. "He is passionate about every aspect of the school—he is even passionate about the guy who waters the lawn . . . but Macchia deflects any talk that he is at the center of the improvement. All important decisions made at the school, he said, are the results of collaborative effort. The teachers, he said, are the ones 'in the trenches—I'm the cheerleader who facilitates'" (187).

Teachers must be committed to student success: "Teachers don't stop teaching just because the state tests have been given for the year. And they use the verb 'to teach' properly. That is, they do not say what many teachers around the country say: 'I taught it, but the kids didn't get it.' Although common, this formulation actually makes no sense. If I were to say, 'I taught my child to ride a bike,' you would expect that my child could ride a bike. She might be a bit shaky, but she should be able to pedal and balance at the same time. If she can't do that, you would expect me to say something like, 'I tried to teach my child to ride a bike'" (216).

6. Medina Study of Programs for High-Risk Youth

Enrique Medina conducted rigorous multivariate analyses, combined with teacher interviews, to examine successful programs for high-risk youth. Over a considerable period of time, he painstakingly assembled a data file drawing together information about alterative programs for at-risk students in the Pomona, California, Unified School District. His study yielded two main conclusions:

> 1. California standardized test scores in language arts increased more at the Village Academy High School (VAHS), an innovative program characterized by small classes, than in either the regular high school or a standard intervention program. . . .
> 2. Teachers who spent time building relationships with their students especially increased the academic performance of students.[15]

The VAHS curriculum involves a focus on technology (employing technology in instruction, preparing for technical certification), real-world educational experiences (internships with businesses and nonprofits and project-based learning), and accelerated learning opportunities (30). Students at VAHS are simultaneously enrolled in high school courses and community college courses. When they have completed high school and received their diplomas, they can earn as much as students who receive a two-year associate's degree from a nearby community college.

Employing an analysis of covariance, Dr. Medina reported statistically significant differences between the VAHS program, the regular high school program, and the intervention program on a standardized test in English Language Arts (ELA). Students in the comprehensive high schools improved by 7.64 percent on the scaled ELA in the ninth grade. Students in a conventional intervention program improved by 3.66 points. Students in VAHS improved by 10.88 points (80).

7. Study of Bennett Woods Elementary School

Finally, let's examine one highly successful school from a comfortable, middle-class community. Note that success looks the same, no matter the community. Michael Pressley and associates studied such a successful elementary school. The Bennett Woods Elementary School outperformed other schools on Michigan State reading and writing tests. Yet when the authors searched the literature for an explanation as to why this school, whose students were relatively advantaged, was the highest-performing school in its area, they found little to help them, especially at the elementary school level. In summarizing what they could find in the literature, the authors concluded that effective elementary teachers tend to "devote much of their class time to academic activity, engaging most students consistently in activities that require them to think as they read, write, and discuss. Effective teachers do explicit teaching (and re-teaching as needed) of skills, and this teaching includes modeling and explaining skills, followed by guided student practice. That is, effective teachers show a strong balancing of skills instruction and holistic reading and writing activities."[16]

In reporting about Bennett Woods Elementary School, Pressley and his colleagues noted that books were freely available and that the school's library was well stocked. Teachers cited professional development as critical to their success, and they worked hard to prepare their students for high-stakes tests. The school hired additional reading specialists and continuously encouraged and

A STORY FROM THE AMERICAN REVOLUTION

A stranger on horseback in civilian clothes came upon a group of army men who were rebuilding a fort. All but one of the men carried materials and put them in place on the structure. The other man barked orders. The man on horseback asked why he was not helping with the fort. The other man replied, "They are privates and I'm a corporal." The man on horseback asked if he could help the project. The other man agreed. The man on horseback then dismounted and spent several hours carrying materials and helping rebuild the fort. Then he got back on his horse, turned to the corporal, and said, "If you ever need help in the future, your commander-in-chief will be glad to help you again."*

George Washington understood the meaning of leadership.

*Robert Gates, U.S. Secretary of Defense, Commencement Address, West Point Military Academy, May 23, 2009, reported by Blackanthem.com military news.

facilitated parent participation: "The norm at every grade level was for students to be reading several books at a time, consistent with the perspective that voluminous reading and exposure to literature positively affect reading. . . . In the primary grades, each student had a book bin of 10 or more books currently being read at school" (229).

There was a consistent emphasis on learning new vocabulary words and the teachers demonstrated how this was vital to understanding the meaning of the message in what they were reading. The teachers frequently assessed their students' knowledge.

Leadership in School Districts

One of the nation's most successful school superintendents, Arlene Ackerman has been an unusually effective leader in four cities. She was deputy superintendent in charge of academic issues in Seattle. She was superintendent of the Washington, DC, and San Francisco school systems. Recently she accepted an appointment as superintendent of the Philadelphia School District. Student achievement scores in each of her school districts increased significantly under her leadership. The Broad Foundation, which focuses on preparing future educational leaders, has recognized her national success and employs her as a mentor of superintendents.

Dr. Ackerman insists on rigorous measurement of student achievement, equity for all socioeconomic and ethnic groups, fiscal responsibility, and linking school funding to achievement. In San Francisco she introduced site-based budgeting, which allows each school to set its own budget, and an open-enrollment system in which parents can choose their children's schools.[17] She also put in place an innovative school funding algorithm called the "weighted student formula," originally used in Edmonton, Alberta, in 1976. The weighted student formula "allows money to follow students to the schools they choose while guaranteeing that schools with harder-to-educate kids (low-income students, language learners, low achievers) get more funds." As used in San Francisco, "the weighted student formula gives each school a foundation allocation that covers the cost of a principal's salary and a clerk's salary. The rest of each school's budget is allocated on a per student basis. There is a base amount for the 'average student,' with additional money assigned based on individual student characteristics: grade level, English language skills, socioeconomic status, and special education needs. These weights are assigned as a percentage of the base funding."[18]

Dr. Ackerman emphasizes test scores, trend data, and performance targets. In 2006, reflecting on her San Francisco experience, she said that these emphases and the changes instituted in the school system resulted in

> five consecutive years of academic improvement for all groups of students at every level. I mean all groups—even special ed.
>
> When I first came to the district, the African-American students' achievement was going backwards. We reversed that. The last two years we have been the highest-performing large urban school district in California. This last year we were up for the Broad Prize as one of the five top urban school systems in the country. I'd say that's pretty good.[19]

In 2010 Dr. Ackerman won the Richard C. Green Award, given to the best urban school leader in the country, from the Council of Great City Schools. Michael Casserly, the executive director of the council, said Dr. Ackerman "is most worthy of the nation's highest individual award in urban education."[20]

Leadership in Colleges or Universities

Next, consider leadership at the college or university level. Successfully leading a college or a university requires consummate skills as a manager, an academic, a visionary, a spokesperson, and a fund raiser. Successfully launching a new college requires even greater skills.

The founding president of Harvey Mudd College is one of the stellar success stories in American higher education. In contrast, the founding president of Harvard College was a disaster.

Harvey Mudd College

Joseph Platt has had at least three highly successful careers: as a physicist, as a college president, and in retirement. A leading young physicist in the United States at the start of World War II, Platt worked at the radiation laboratory at the Massachusetts Institute of Technology (MIT). He served as a technical advisor in both the European and Pacific theaters of the war. After the war, he continued his career in the University of Rochester physics department.

In 1955, Joe Platt received a phone call that changed his life. He was invited to become founding president of Harvey Mudd College, an engineering school in the complex of Claremont Colleges in California. He, and his wife, Jean, a mathematician, accepted the challenge.

When Platt arrived in Claremont, Harvey Mudd College was nothing but an idea, a plot of ground, and some start-up funds. He brought his keen intelligence, extraordinary administrative skills, good humor, and energy to the task of building a college from the ground up. Gene Hotchkiss comments:

> Joe recruited his first faculty not with promises of prestige, advancement and ample salary (as do most college presidents), but with his persuasive enthusiasm and by sharing his excitement in a great educational venture to be undertaken. Those who signed on in the early years found his courage contagious. With his leadership, they dared to build a new and different curriculum; to set demanding standards for themselves and their students; and to foster scientific inquiry in an atmosphere of academic freedom. With his leadership, they recruited remarkable students, also adventurers, who dared to enroll in an unknown college. In a short time, again with his leadership, the excitement of this new college spread to those corporations and foundation leaders whose support would ensure the college's future.[21]

Platt's achievements, and the rise of Harvey Mudd College, may be unique in the history of American higher education. Within 25 years, Harvey Mudd was thriving, successful, and ranked among the top engineering colleges in the United States—up there with MIT and Cal Tech. Harvey Mudd is known as a school specializing in science and engineering that emphasizes the humanities more than its peer institutions.

One of this engineering school's founding board members has written: "The early history of Harvey Mudd College is as much the story of an individual as it is of the years of a precocious institution. Dr. Joseph B. Platt served as midwife and parent to the college as it experienced the pleasures and pains of . . . rapid growth. . . . The college of today, recognized nationally as a 'producer' of future engineering and science Ph.D. graduates, owes its intellectual wealth and its prominence, more than any other, to this remarkable leader."[22]

Harvard College

Contrast Joe Platt's success with the tale of Nathaniel Eaton, founding president of Harvard. Harvard takes great pride in its unparalleled excellence and in its storied history as the first institution of higher education in America. But it didn't start well.

Nathaniel Eaton came to Harvard in 1636 with impeccable credentials. He held a graduate degree in theology from Cambridge University. Expectations couldn't have been higher. Within a year, those expectations had been completely dashed. Mr. Eaton was on trial in a Cambridge court for having lost his temper and hitting a subordinate over the head with a two-by-four. In addition to this incident and related concerns about what might be called an insensitive management style, there were other accusations from the campus community. For example, his wife was accused of watering down the beer served to the undergraduates. Mr. Eaton was found guilty of the assault charge and was fired as president of Harvard. He died in a debtor's prison. Harvard was temporarily closed.

Reflecting on the choice of Nathaniel Eaton as first head of the college, Charles Wagner comments: "Undoubtedly his studies at Franeker under the great William Ames, and the fact that he published under Ames a Latin paper on the nature of the Sabbath, must have influenced his appointment strongly. This was one time that the Harvard dictum 'publish or perish' backfired hard, and early."[23]

Harvard reopened after a year with a new, dynamic president, Henry Dunster, also a young graduate of the University of Cambridge. According to historian Samuel Eliot Morison, although Dunster was "the youngest in the long line of Harvard presidents, he proved to be one of the greatest. But for his faith, courage, and indomitable energy, Harvard College might have become a mere boarding school, or have closed altogether during the gloomy depression that settled over New England before he had been in office a year."[24]

Dunster was a determined fundraiser for the new college, even once selling a shipload of Brazilian wood to raise funds for additional real estate purchases. Morison notes that because of his genius for organization, "the curriculum, the forms, and the institutions established under his presidency long outlasted his time, and even his century. Harvard University grew out of the Liberal Arts

College as Dunster left it; and the Charter of 1650 that he obtained, and in all probability drafted, still serves as constitution of the modern University."[25] After the stumbling start under Nathaniel Eaton's leadership, Harvard went on to become one of the world's great universities, thanks in great part to Henry Dunster.

Evaluation and Assessment

Effective leadership and organizational change require that new initiatives be evaluated. We cannot reform education if we do not apply the most rigorous quality standards. We cannot know for certain that pedagogical innovations are helping students unless we perform hard-nosed assessments. Evaluation is a relatively new discipline with its own norms, standards, and techniques. Educators must learn these techniques. Far too often, sweeping changes are implemented in our schools or colleges, often at great cost, without careful monitoring or evaluation.

In the second half of the 20th century, American social scientists developed a new technology with the potential to revolutionize management and planning by governments, educational institutions, and private corporations. This technology is not hardware or gadgets but rather a collection of ideas, techniques, and methods. Taken together, they define the emerging field of evaluation—a new discipline devoted entirely to assessing the success or failure of organizational programs. To paraphrase the psychologist Hermann Ebbinghaus about his field, "evaluation has a long past but a short history." Since the beginning of time, managers have needed to assess the performance of people or organizations under their control. But evaluation as a discipline has a short history; only recently has the power of modern social science been applied to such judgments.

Consider a national program at the National Science Foundation or the Department of Education. This program is intended to fund a number of local projects. What are the criteria for evaluation?[26] Below are five recommendations for program evaluation. I describe recommendations 4 and 5 in greater detail so as to give a clear picture of what is included in an effective evaluation process and product.

1. **Require an Evaluation.** Require that evaluations be conducted of *each* funded project and, more generally, of the national program. Too many programs and projects are never evaluated. This is a nontrivial problem. STEM education is too important to the nation's vital interests to be left to chance, good will, and anecdotal recollections.

2. **Include Mixed Methods.** Insist that both (a) quantitative methods based on hard data, including multivariate statistical analyses, and (b) probing qualitative methods (e.g., interviews, focus groups, and narratives) be employed. Quantitative methods can profile trends over time and can statistically isolate program or project effects. Qualitative methods provide deep information on the psychological, cognitive, and sociological components of success.
3. **Avoid Testimonials.** Be wary of "evaluations" that consist of testimonials from project directors about how valuable the funded efforts were. Anecdotes from participants are not particularly helpful, and they are certainly not objective.
4. **Include Process and Outcome Evaluation.** In the literature these often are labeled "formative," or process, and "summative," or outcome. Both should be part of any evaluation plan.

   Process evaluation examines the ongoing implementation of the funded activities and provides periodic feedback to the directors of the effort. In part, this means monitoring whether funds are being applied as originally proposed. For example, have proposed summer internships been established? Process evaluation involves assessing and providing feedback on which "treatments" or interventions seem to be working and which interventions need adjustments. Spring training, in which athletes and game strategies are assessed prior to the major league baseball season, is a kind of process evaluation.

   Outcome evaluation provides information on whether measurable objectives have been achieved. More to the point, these assessments isolate and identify program or project effects. There are several key steps in outcome evaluation.

   a. *Specify measurable outcomes.* Each program or project should have explicit goals that can be translated into measurable objectives. It is astounding how many well-intentioned, hard-working individuals lead organizations but cannot articulate clearly what the organization's goals are. Statements such as "improving the quality of science education in undergraduate colleges" are not specific enough. This is not meant to be mindless reduction of complex human activities to trivial numerical indicators; rather, it is about knowing, and being able to state, what the purpose of the funding is. Specify how each outcome was measured. Programs should demand clear tests of achievement as one key measure.

b. *Measure abstract constructs with clarity.* There are unclear and inconsistent definitions of "underrepresented minority students" in the field and in the literature—for example, how are Pacific Islanders classified? Defining what constitutes or defines a university can vary. Is the University of Maryland defined and measured as all campuses combined, as College Park only, or as each campus separately? Disciplines can be categorized differently by different funding programs and institutions: is astronomy listed independently or combined with physics? Clear, simple descriptions and measurements of such abstract concepts are critical.

c. *Include baseline data.* Key variables about program participants should be gathered before any intervention. This information gathering is critical to judging later success or failure. This may seem obvious, but many evaluation reports estimate or assume pretest conditions because they failed to measure them.

d. *Measure the intervention.* There should be no ambiguity in the evaluation data about whether an individual was a program participant. Students move into and out of middle schools, into and out of college majors, into and out of program funding, and the like. This movement must be tracked and described carefully. Any ambiguity can undermine the evaluation.

e. *Place outcome data in context.* Presenting outcome statistics alone is insufficient. Outcome data should be compared to the baseline data to demonstrate value-added trends. Ideally, such growth or value-added changes (i.e., pre- to post-intervention) should be compared with national trends. For example, in Chapter 6, I present data showing that a Houston program had almost doubled the number of underrepresented minority students receiving STEM bachelor's degrees in five years. This consortium growth rate was far greater than the national growth in minority student STEM degrees during the same time period.

5. **Include Advanced Evaluation Components.** Three components are necessary for a more complete and more valuable comprehensive evaluation: a comparison group, linkage of activities to value-added changes, and an examination of the program's return. Although these components may go beyond what is reasonable or feasible for some evaluations, such advanced evaluation techniques can help pinpoint the areas of strength and weakness of the program being studied.

Employ a comparison group. Choose a comparison group of similar students and gather pre- and post-data as a benchmark. In an ideal setting, the treatment and control groups would be chosen randomly in an experimental design. But doing so is rarely practical (program and project participants are usually chosen for a reason, not randomly). Still, it often is possible to select a comparison group that closely matches the recipient group, in what is called a "quasi-experimental" design.

Link program activities to value-added changes. When pre- to post-intervention changes are uncovered, conduct statistical tests (such as analysis of covariance) to reveal whether these changes can be attributed to the intervention.

Examine the return on investment. When program effects have been isolated through multivariate analyses, relate them to the project funding. How much did it cost to achieve these effects? Consider also whether money was saved by the program; for example, by moving students through college faster.

In summary, each program or project evaluation should

- include both qualitative methods (e.g., interviews) and quantitative methods (e.g., statistical analysis),
- have both a process and an outcome component,
- be based on clear, unambiguous definitions of the treatment and desired outcomes, and
- employ pre-test and post-test data to isolate the value-added changes.

Evaluation in education is difficult because there is no consensus about what constitutes quality—research quality, institutional quality, or teaching quality. But this problem is not unique to academia, as Robert Pirsig observed in *Zen and the Art of Motorcycle Maintenance:*

> Quality . . . you know what it is, yet you don't know what it is. But that's self-contradictory. But some things are better than others, that is, they have more quality. But when you try to say what the quality is, apart from the things that have it, it all goes poof! There's nothing to talk about. But if you can't say what Quality is, how do you know what it is, or how do you know that it even exists? If no one knows what it is, then for all practical purposes it doesn't exist at all. But for all practical purposes it really does exist. What else are the grades based on? Why else would people pay fortunes for some things and throw

> others in the trash pile? Obviously some things are better than others . . . but what's the "betterness"? . . . So round and round you go, spinning mental wheels and nowhere finding anyplace to get traction. What the hell is Quality? What is it?[27]

Evaluation of teaching in higher education has received less emphasis in the literature than evaluation of research, partly because of an emphasis on research in academic life (on the tenure track, research is valued more than teaching) and partly because of the difficulties in assessing teaching performance. The usual method has been through student evaluation of professors. Student evaluations run the risk of turning undergraduate education into a popularity contest, with professors giving easy grades in an effort to improve their ratings among students. A more serious problem from an analytic point of view is that findings about professors cannot be compared across institutions. It is difficult, if not impossible, to compare the quality of teaching in different schools.

Rigorous Measurement of Outcomes

As an evaluation researcher, I have been astounded to discover how many intelligent people working hard in organizations cannot articulate the goals of their organization. Often, it is an external evaluator who first forces articulation of objectives. One cannot evaluate whether an organization has achieved its goals until those goals are stated. Occasionally, some program directors find questions about goals to be threatening.

When I worked at the Rand Corporation, a colleague and I conducted a (classified) study of officer training, or accession, for the Office of the Secretary of Defense. The question was whether the best officers came from the service academies, ROTC, or officer training schools. We were particularly interested in whether the quality of new officers could be improved by manipulating the fellowship aid provided to undergraduates through ROTC. At a meeting in the Pentagon, I suggested that an additional study was in order. If we were to assess what mechanisms created the best officers, we needed a clear definition of success as an officer. We needed to do a study of who was promoted, and why. (There were rumors of "ring knocking ceremonies"—i.e., that graduates of the service academies were in the inner circle and were more likely to be promoted.) My suggestion for a rigorous examination of who got promoted and why was greeted with disbelief and icy stares. There was absolutely no enthusiasm for my suggested additional study.

In any study, data must be interpreted carefully. Depending on the interpretation, the same statistic can be viewed as indicating either success or failure. The correct interpretation always involves careful consideration of the context and meaning of the measure.

Consider college attrition rates. (Attrition rates are commonly defined as the percentage of students who drop out before graduating from an institution.) Some leading institutions boast about their challenging curriculum and point to a high attrition rate as evidence. College lore is full of stories of speakers saying to a group of entering freshmen: "Look to your right. Look to your left. One of the three of you will be gone by the end of the year." Conversely, high attrition rates have been used to argue that institutions are substandard.

Hospital ratings are similar to college attrition rates. Sometimes hospitals are ranked by mortality rate—that is, the percentage of patients who die while in the hospital. High mortality rates are often interpreted to be an indication of a poor hospital. But the reverse may be true. Hospitals with the greatest expertise in treating a certain illness or condition may receive the sickest patients, patients who may be close to death.

Statistical data such as mortality rates must be interpreted in context. Consider the apocryphal story of a sociologist who approaches an elderly Maine farmer. He says, "I'm a sociologist with the public health service and I'd like to ask you some questions. We're trying to determine the mortality rate in this part of Maine." The farmer replies, "Well, son, as best I can tell, the mortality rate around here is about one death per person."

Assessment and No Child Left Behind

Accurate assessment of student achievement is vital for strengthening STEM education in America. Our current era of high-stakes testing is perhaps epitomized in the federal No Child Left Behind (NCLB) Act. Some components of this act have resulted in improvements. For example, statewide testing can correct for teacher grading biases, and disaggregated student testing data allow decision makers to identify schools with overall high achievement levels that are shortchanging students of a particular ethnic group or students from impoverished backgrounds.

But national testing initiatives like NCLB have three major problems:

1. There is too much testing. Excessive testing steals time from the instructional program and significantly increases student anxiety levels. Both impede student learning.

2. Virtually all testing is done with standardized, multiple-choice exams, although we know that such testing only scratches the surface of the assessment of learning.
3. Norm-referenced tests, which score on the curve, are often used. Instead, we should assess whether a student has mastered the material, regardless of what his or her peers know, through criterion-based tests.

William G. Spady, a nationally respected scholar, has been articulating a vision of educational reform for decades. In a widely discussed article, "The Paradigm Trap: Getting Beyond No Child Left Behind Will Mean Changing Our 19th-Century, Closed-System Mind-Set," he observed: "When Americans recognized the hard realities of the Information Age 25 years ago, they faced a profound challenge: become future-focused and change, or stay the course and become obsolete. To survive, U.S. business chose the former. To provide continuity with the past and not rock the boat, education's power brokers in contrast chose the familiar, tried-and-true route."[28]

As a metaphor to illustrate these problems with NCLB, imagine that the federal government and the states launched a new program to evaluate and improve hospitals. Suppose they decided to conduct rigorous assessment of the health status of patients released from each hospital.

- Suppose the testing program assessed only blood pressure, ignoring other vital indicators of health status. Serious health evaluations require a comprehensive review of health in general, not just a single value of a single variable.
- Suppose the evaluation required measuring patients' blood pressure every half hour. This repeated testing would add little to our assessment of a hospital, but it might prevent patients from engaging in the very activities—for example, aerobic exercise—that could help lower their blood pressure.
- Suppose hospitals and patients were assessed not against a nationally recognized standard (e.g., 120/80) but in comparison with other patients. What if many of the comparison patients had high blood pressure, partly due to the effect of poverty on their diet? A hospital with a preponderance of such patients might wind up erroneously being rated superior.
- Suppose the hospitals that performed poorly on the blood pressure tests were told they had to improve or risk heavy federal sanctions, but were given no resources, such as blood pressure medication or educational materials, to change the situation.

- Suppose the best health professionals left hospitals for private practice because the constant focus on assessing blood pressure dramatically interfered with their ability to do their jobs.

This is what we are doing to our schools and to our students. In my opinion, tests and testing programs should

- use criterion referenced tests only;
- sharply reduce the number of school days and hours devoted to testing;
- incorporate more testing modalities, beyond multiple-choice exams, even if that means more work for those conducting the tests;
- insist that the teacher's holistic judgment be weighted more heavily than the standardized test results in a given student's assessment.

The Impact of Evaluation and Assessment

Evaluation of programs to improve STEM education is critical. But even a sophisticated, rigorous, meticulous evaluation is rendered meaningless if leaders don't use the information to change the program. All too often, there is a gap between research and evaluation, on the one hand, and policy and practice, on the other.

If educational reform is to be effective, educators and political leaders need to learn a common language and depart from jargon. For example, educational researchers (hypothetically) write articles that might say: "In the hierarchical multiple regression analysis, the beta for the treatment effect approached significance. If we had access to a larger sample, thus avoiding a possible Type II error, or had measures of additional covariates, the beta might have been statistically significant. Nonetheless, the regression findings are of great heuristic value."

Translation? "The program didn't work. Maybe we did the evaluation study wrong. If we're lucky, we can find something useful here."

Politicians say things like this in their speeches: "The problems in our schools really aren't that complicated. All we need to do is return to the basics." Translation: "I've got a good job, earn good money, and I went to a small town public school during the Eisenhower administration. If we would just return to the 1950s, everything would be fine."

Politicians believe many educators can't speak English and deal directly with problems. They are right. Jargon serves no purpose except to make professors feel important. It obscures the important messages the researchers are trying to deliver. Educators believe many government officials don't understand the complexity of the social, educational, and economic problems that plague schools.

They are right. Politicians have a low tolerance for nuance. Scholars have learned a lot about what works in schools. But they have failed to communicate this to the public and to our political leaders.

Moreover, neither professors nor politicians seem to have a sense of history. Here is a quotation from a California governor: "We find ourselves threatened by hordes of . . . immigrants, who have already begun to flock into our country and whose progress we cannot arrest."[29] No, this is not former Governor Pete Wilson talking about immigration from Mexico. In fact, these words were spoken in 1846 by Pio Pico, the last Mexican governor of California, referring to "Yankee" immigrants.

The recent debates about bilingual education in California and elsewhere have been conducted as though Americans had always been taught in English and only the recent influx of immigrants from Mexico and Asia forced us to consider instruction in other languages. Yet, in the early days of this republic many schools, particularly in Pennsylvania, carried out instruction entirely in German. One expert estimated that one million pupils attended public bilingual schools during the 1800s, when the U.S. population was much smaller. In the Third Congress, two congressional committees debated whether to print all federal laws in German as well as in English.

If evaluation and assessment are to be effective, educators and researchers must learn how to communicate with school boards, state officials, boards of trustees, and Congress, as Peter Drucker learned when he was asked to testify before a Senate committee on supplies and materiel chaired by a little-known senator from Missouri, Harry Truman. After the session, Truman invited Drucker back to his office, where the senator offered him a tumbler full of bourbon. Truman asked him if this had been his first time testifying before Congress and Drucker replied that it had. Truman complimented him and said that he had done an excellent job. He then asked if Drucker would mind if he gave him some advice. Drucker said he would welcome advice. Truman said, "When you are presenting to United States senators, do not use sophisticated mathematical concepts . . ." Then there was a long pause: ". . . like percentages."[30]

CHAPTER 4

Top-Notch Teachers

> The last change in teaching was 550 years ago in the 15th century when the printed book came in. We are still teaching in exactly the same way.
>
> Peter Drucker

No matter how many task forces are convened, how many curriculum projects are funded, or how many high-stakes tests are given, successful education ultimately comes down to the interaction and communication between a teacher and his students.

Therefore, the surest way to improve STEM education in America is to improve teaching. We need to

- attract the best college graduates to teaching,
- provide them with meaningful education and training,
- keep them from leaving the position in despair,
- provide professional development for current teachers,
- respect these professionals to whom we entrust the education of our children, and
- pay them appropriately.

This chapter describes findings and conclusions from the latest research about teachers and profiles some extraordinary programs that are making a difference. It focuses on how to recruit better teachers and on what effective teachers do. Unfortunately, many American students do not currently have access to outstanding teachers, teachers with expertise in the fields they teach.

Teacher Quality and National Achievement

The research literature addresses teacher quality mostly at the general level of all subjects, not just STEM fields. Because most of this literature also applies to STEM, we begin by considering the teacher quality problem, both in general and in STEM.

Motoko Akiba, Gerald LeTendre, and Jay Scribner studied teacher quality and national achievement in 46 countries using data from the 2003 Trends in International Mathematics and Science Study (TIMSS) research study. They concluded that although the quality of teachers in America is about the same as in other countries, an opportunity gap exists between students depending on their socioeconomic status. Between poor students and privileged ones the access gulf is "among the largest in the world." Their research uncovered "empirical, cross-national evidence of the importance of investing in teacher quality for improving national achievement."[1]

Akiba and colleagues also present data about teacher qualifications in each of the 46 nations that participated in the 2003 TIMSS study. I was particularly interested in the likelihood that a student in each country would be taught mathematics by someone who had been a math major in college. Frequently, students in secondary school fail to learn math because their teachers simply do not understand the subject. Table 4.1 presents data which show that the United States ranked 41 out of 46 countries. The United States' ranking is troubling. Students in Lithuania, Syria, Tunisia, Iran, Botswana, Jordan, Ghana, and the

TABLE 4.1
Global rankings and percentages of math teachers with math degrees

1. Latvia	98.1%	24. Israel	74.6
2. Cyprus	97.7	25. Morocco	73.1
3. Bulgaria	96.6	26. Jordan	72.3
4. Romania	96.5	27. South Africa	68.1
5. Flemish Belgium	95.9	28. Sweden	66.7
6. Russia	95.6	29. Philippines	62.4
7. Serbia	95.6	30. Hong Kong	61.9
8. Saudi Arabia	93.3	31. Australia	61.5
9. Lithuania	92.3	32. Macedonia	60.0
10. Moldova	87.5	33. Slovak Republic	59.5
11. Singapore	85.7	34. Indonesia	58.8
12. Egypt	85.1	35. Ghana	57.4
13. Armenia	84.6	36. Chile	52.9
14. Syria	84.1	37. Palestine	51.2
15. Tunisia	82.7	38. New Zealand	49.9
16. Slovenia	80.9	39. Netherlands	49.4
17. Japan	80.7	40. Bahrain	48.9
18. Estonia	79.9	**41. United States**	**47.3**
19. Taiwan	79.7	42. Malaysia	45.9
20. Iran	77.8	43. Hungary	40.9
21. Botswana	77.1	44. Korea	40.4
22. Scotland	75.9	45. Norway	37.4
23. England	75.6	46. Italy	20.8

Source: Data from Motoko Akiba, Gerald K. LeTendre, and Jay P. Scribner, "Teacher Quality, Opportunity Gap, and National Achievement in 46 Countries," in *Educational Researcher* 36, no. 7 (2007): 369.

West Bank and Gaza are more likely to have a teacher with a math degree than are students in America!

One Program That Works: CGU Teacher Education

One particularly useful model for improving teaching skills is the on-site graduate mentoring provided by the Teacher Education Internship Program (TEIP) at my institution, the Claremont Graduate University (CGU). College graduates enter the program with an intensive summer of instruction (all day, every day) in pedagogy and classroom management that includes some practice teaching. (Many California schools are on a year-round schedule and, thus, there are ample opportunities for practice teaching placements. A few students gain their practical teaching experience as undergraduates in summer school.) In September, students become fully responsible and fully paid teachers of record in a classroom as an intern. They receive intensive guidance and feedback from a Claremont faculty advisor who visits the classroom and meets with the intern frequently during fall and spring graduate coursework.

After the first year, program participants enroll in another intensive summer at CGU, after which the students have completed the requirements for a teaching credential. After taking some additional PhD-level electives, they also will have completed the requirements for a master's degree in teaching. Thus, 15 months after earning bachelor's degrees, the students have acquired a teaching credential, a master's degree, and a paying job. The TEIP faculty have worked closely with a number of proximate school districts to facilitate placement of interns in teaching positions; often the intern remains in that position following receipt of the master's degree.

In the 2009–10 cohort of teacher education interns at CGU, 30 percent were white, 32 percent were Latino, 17 percent were Asian or Asian American, 13 percent were African American, and 8 percent were some other ethnicity. Students had earned their bachelor's degrees at a wide range of institutions, including the University of California campuses at Los Angeles, Irvine, and Berkeley; the University of Southern California; Claremont McKenna College; and Harvey Mudd College. Of the 100 students, 40 students spoke languages other than English, including Spanish, Vietnamese, Korean, Mandarin, French, and German.

In 2008, Lisa Loop, the interim director of CGU's TEIP, surveyed alumni who had graduated at least five years earlier. She found that more than 90 percent were still teachers. This high retention rate is significant, as many teachers leave the profession (estimates range as high as 50 percent).[2] Teacher attrition continues to be a major reason for the shortage of excellent teachers in the United States. In this CGU sample, however, only 4.17 percent had left teaching five years after graduation. Another 5 percent were not teaching because they were on family leave, but they had no intention of leaving the profession.

Loop conducted an advanced statistical analysis, hierarchical regression, to assess the three factors associated with teachers staying in the profession:

1. pre-program student characteristics (e.g., college grade point average)
2. program strategies and activities
3. the post-degree work environment

She found that almost all the variation in staying versus attrition could be attributed to the CGU program. She concluded that "teacher preparation variables have the strongest predictive power for both teacher retention and success. School support variables also showed positive effects on teacher retention and success. Finally, in this study, pre-graduate school variables showed no predictive power in teacher retention or success."[3]

An earlier follow-up study of TEIP graduates in the 1990s also reported a retention rate exceeding 90 percent. The internship model pioneered by CGU now has been adopted by a number of universities in California. (This model has been under some stress during the current recession, as some districts are cutting back on, or eliminating, new hires.)

Fewer Tests, More Teachers

Policymakers have admirable goals, and the federal government and the states have mandated many high-stakes tests. But they focus on diagnosis, not cure. We already know what the problem is. It's time to fix it.

Participation in the global economy demands mastery of technical subjects. Research repeatedly shows that American students stumble in math and science in middle school and fall down badly in high school. Often the gatekeeper course is algebra.

Suppose that, to confront the childhood obesity epidemic, the state of California mandated that every high school graduate compete competitively in all decathlon events, including the pole vault. But suppose no additional coaches were hired and current coaches received no additional training. The national approach to improving student performance in math has been similar for the past decade.

- There is nothing wrong with adding a test to the curriculum.
- There is something wrong with *only* adding a test.
- While the politicians pontificate, America's students struggle to learn mathematics and science against staggering odds.

Fortunately, bold and innovative leaders, including leaders from the private sector, are trying to bring fundamental change to the mathematics education infrastructure. Two new funding programs, Math for America and the National Science Foundation Noyce Scholarships, described next, hold great promise.[4]

Math for America

One person can make a difference. Billionaire Jim Simons, a successful investor and a mathematics professor, was concerned about the weak math and science skills of U.S. students, so he decided to do something about it. With personal seed money (about $50 million!) and additional funding from others, Simons launched Math for America with a pilot program in New York City. The Math for America program recruits outstanding college graduates who have studied advanced mathematics and fully pays for their graduate education as teachers. The program also gives students a substantial salary bonus for six years once they start teaching and provides professional development during that period.

Math for America is expanding geographically. I serve on the steering committee for a collaborative effort by three institutions in the Los Angeles area: CGU, Harvey Mudd College, and the University of Southern California (USC). Harvey Mudd president Maria Klawe calls the program "a great resource for the hard-working math teachers already in the schools as well as the new teachers it trains and supports."[5] In 2008, the collaborative was launched with a cohort of 12 impressive future teachers. Half attended USC and half CGU. The collaboration among the three institutions has been collegial, synergistic, and successful. Math for America Los Angeles is subject to a mixed-methods rigorous evaluation, which includes assessing how new teachers can help improve student achievement test scores. Marla Mattenson is one student from the first cohort, and I tell her story here.

In addition to Mattenson other graduates of the program are also enthusiastic about the results. In Math for America's New York program, one participating teacher, Nic Vitale, a math and science teacher, reported that he was

A MATH FOR AMERICA SUCCESS STORY

Marla Mattenson graduated from UCLA with a BS in mathematics and applied science. She was in the first group to enroll in Math for America at CGU.

Marla believes she was called to teaching, which she views as the ultimate form of service. She has also helped women through labor and delivery as a doula, practiced massage therapy, and created and taught workshops to address math anxiety.

In addition to teaching, she feels her job is to inspire, motivate, and guide students. She believes that when a child understands how to solve a problem from different perspectives, a sense of wonder can ensue.

Marla sees education as a tool that helps kids understand their patterns of stress, with the patient teacher as the lens. Every child—regardless of race, ethnicity, or economics—deserves to experience a thriving learning environment. This self-awareness in the classroom inspires children to grow into reflective, deeply compassionate adults.

Marla was appreciative of the fellowship because the training she received will allow her to play an integral role in building a strong community of math teachers who will help raise the quality of math education in America.

After her first year as an intern mathematics teacher, Marla was asked to chair the mathematics department at Bernstein High School in Los Angeles.

starting to get "a little worn out" after eight years in New York City's classrooms. "A lot of people were asking me, 'Why don't you move to the suburbs?'" he says. Then he heard about Math for America's master teacher program. Soon, he was attending professional development programs, mentoring young teachers . . . and cashing his bonus checks. "Now, nobody asks me why I'm here," Vitale says.

Why would they? He's being paid like the professional he is.[6]

Jim Simons has a vision through which the federal government or the private sector could leverage the Math for America model to transform teaching in America. He notes that there are roughly 250,000 mathematics teachers in the United States today. Consider a program in which the government agrees to augment the salaries of the most effective math teachers, say 100,000 teachers (i.e., the top 40%), by $20,000 per year. The total cost of this effort would be $2 billion per year.[7]

Consider the effect on mathematics teaching if those who were most effective at enhancing student achievement were awarded such salary bonuses.

While two billion dollars sounds like a considerable sum to you and to me, it would represent a tiny percentage of the annual federal budget.

Obviously, defining who is an "effective teacher" in a fair manner will be a challenge. But this implementation hurdle is not insurmountable. And a federal program like this could elevate teaching to the high status this profession holds in many other countries. It could draw excellent new teachers into the profession, and it could dissuade many from leaving the profession.

Noyce Scholarships

The National Science Foundation (NSF) is drawing outstanding college graduates in math and science across the country into teaching through the Robert Noyce Teacher Scholarship program. Originally, NSF Noyce Scholarship support contributed $10,000 (now up to $20,000 under a new subprogram) to a student's pursuit of a teaching credential and a master's degree in return for a commitment to teach for at least two years in a high-need district. The program also supports undergraduates preparing to become teachers who are majoring in mathematics or a science discipline (up to $7500). In addition, under some Noyce grants, the new teachers receive a salary supplement during their first two years as a teacher (up to $10,000 per year). Finally, Noyce also funds some Math for America students.

Eric Edens, a student at CGU, was awarded a Noyce grant and was assigned to teach in a California public school as an intern. He and his students developed an intracampus Web site that was featured in *Edutopia*, a George Lucas Educational Foundation publication. The model for the site was suggested by a creative student, a former gang member who initially had not been interested in the project.

Another Noyce fellow at CGU, Andrea Cuellar, came to the United States after receiving her elementary, secondary, and undergraduate education in Argentina. She teaches chemistry at Baldwin Park High School. When she arrived in the United States, she was appalled to discover that many people assumed she was not bright because English was a new language for her. Similarly, she is shocked when she realizes that many teachers conclude that a student who speaks limited English is not smart. She tells all her students, including the English language learners (ELLs), that they can succeed in chemistry. And her students' test results speak louder than her words. Last year, her ELL students outperformed the district mean for all students, including those who are fluent in English.

What Effective Teachers Do

> I am a teacher at heart, and there are moments in the classroom when I can hardly contain my joy. When my students and I discover uncharted territory to explore, when the pathway out of a thicket opens up before us, when our experience is illumined by the lightning-life of the mind, then teaching is the finest work I know.
>
> Parker Palmer, *The Courage to Teach*[8]

For the past quarter century, social scientists have employed a new tool called meta-analysis for synthesizing and understanding the implications of research. In the past, scholars conducted literature reviews of existing studies on a topic, but they had no mechanism to quantitatively combine results from those studies. Gene Glass developed such a method, based on the "effect size." Effect size takes into account the standard deviation—that is, the amount of variation—in both the intervention and the outcome variables.

Meta-analysis advanced our ability to extract meaning from a set of studies, and the use of effect size allowed us to transcend an obsessive focus on "statistical significance." For decades, publication of a study has depended largely on whether the statistical findings were "significant," usually at the .05 level. Significance calculations essentially tested whether there was an effect. If you found an effect, your study was likely to be published in a peer-reviewed journal.

Years ago, it troubled me that so much journal space was devoted to statistically significant findings that reported differences so small as to be meaningless. In one article, I argued that "the grade inflation that plagues undergraduate education seems to have spread to peer-review evaluations of papers submitted to journals. Tougher methodological and substantive criteria would prune these journals, removing poor work from the academic literature. . . . Too many researchers confuse statistical significance with substantive importance. Statistically significant, but puny, correlations, easy enough to get with a large enough sample, are passed off as important."[9]

Glass and others provided a solution that went beyond simply measuring if we can state with confidence that there was an effect. We can now estimate how great the effect was. And effect sizes allow combining results from studies that employed different samples and somewhat different methods.

Consider the work of John Hattie of the University of Auckland. Like many of us, he studied factors contributing to educational achievement and tried to make sense of the many studies in the literature. Hattie observed that in recent years, educational researchers were publishing meta-analyses—for example, synthesizing the studies that looked at the effects of class size. Hattie decided to synthesize all these meta-analyses in a grand meta-analysis of about 800 meta-analyses. This was a Herculean task. He drew together results and findings from more than 52,000 studies of educational achievement. Collectively, he estimated that these studies involved samples that added up to more than 200 million students (although Hattie is quick to point out that the same student may have participated in more than one study).

Hattie summarized the main conclusions of his meta-analysis, which can be considered signposts of excellence in education:

1. Teachers are among the most powerful influences in learning.
2. Teachers need to be directive, influential, caring, and actively engaged in the passion of teaching and learning.
3. Teachers need to be aware of what each and every student is thinking and knowing, to contrast meaning and meaningful experiences in light of this knowledge, and have proficient knowledge and understanding of their content to provide meaningful and appropriate feedback such that each student moves progressively through the curriculum levels.
4. Teachers need to know the learning intentions and success criteria of their lessons, know how well they are attaining these criteria for all students, and know where to go next in light of the gap between students' current knowledge and understanding, and the success criteria of "Where are you going?" "How are you going?" and "Where to next?"
5. Teachers need to move from the single idea to multiple ideas, and to relate and then extend these ideas such that learners construct and reconstruct knowledge and ideas. It is not the knowledge or ideas, but the learner's construction of the knowledge and these ideas, that is critical.
6. School leaders and teachers need to create school, staffroom, and classroom environments where error is welcomed as a learning opportunity, where discarding incorrect knowledge and understandings is welcomed, and where participants can feel safe to learn, relearn, and explore knowledge and understanding.[10]

Even though the studies that Hattie looked at included structural variables like socioeconomic status, class size, curriculum measures, and many other factors, *the most powerful determinants of educational achievement all involved*

TABLE 4.2
Eight key teacher characteristics leading to optimal teacher-student interactions

1. Non-directivity (activities initiated by students)
2. Empathy
3. Warmth
4. Encouragement of higher-order thinking
5. Encouraging learning
6. Adapting to differences
7. Genuineness
8. Learner-centered beliefs

Source: John Hattie, *Visible Learning: A Synthesis of Over 800 Meta-Analyses Relating to Achievement* (New York: Routledge, 2009).

teacher-student interaction. Table 4.2 lists the eight key teacher-student interaction variables, from highest to lowest effect size, that Hattie found. Hattie adds:

> It is less the content of curricula that is important than the strategies teachers use to implement the curriculum so that students progress upwards through the curricula content. The sharing by teachers of their conceptions about what constitutes progress through the curricula is critical (and this assists in reducing the negative effects of mobility and changing classrooms), as well as ensuring appropriately challenging surface, deep, and conceptual knowledge and understanding. So often changes to curricula are more cosmetic than transformational. . . . Too often there is little attention paid to how to build a common conception of progress across the years studying the curriculum. Teachers need to help students to develop a series of learning strategies that enable them to construct meaning from text, develop understanding from numbers, and learn principles in science.[11]

Hattie concluded that effective teachers had certain qualities that accounted for their success:

> Effective elementary teachers, especially those effective in promoting reading and writing, tend to do the following: They devote much of their class time to academic activity, engaging most students consistently in activities that require them to think as they read, write, and discuss. Effective teachers do explicit teaching (and re-teaching as needed) of skills, and this teaching included modeling and explaining skills, followed by guided student practice. That is, effective teachers show a strong balancing of skills instruction and holistic reading and writing activities. Teacher scaffolding and re-teaching are salient, accounting for a large proportion of such teachers' effort. Effective teachers

> have high expectations and increase the academic demands on their students (i.e., consistently encouraging students to attempt slightly more advanced books and write slightly longer and more complex stories). From the first day of school, effective teachers communicate high expectations for students to self-regulate and take charge of their behavior and academic engagement.[12]

In the above quotations, we see again the importance of leadership, high expectations, and evaluation. They are critical to reforming STEM education and to improving outcomes for our students.

Strategies for Improving Teaching

One explanation for the poor performance of U.S. students, according to Iris Weiss, a STEM education scholar, is the decline in the amount of time teachers devote to hands-on teaching as opposed to lectures. She also observed that while many high school science and mathematics teachers obtained a science degree, their major was often in a different science field than the one in which they presently teach.[13]

As noted in the introduction, the success of the women's movement has contributed, inadvertently, to the decline in the quality of science and mathematics teachers. The educational system has lost the "hidden subsidy" it enjoyed when many bright young women chose teaching because they had few options. I believe this gendered history and sexism explain why teachers do not get the respect they deserve. Even though the gender ratio of teachers has shifted dramatically since the 1950s, some in our society still afford teaching low status as "women's work."

Richard J. Murnane and Randall J. Olsen studied teachers in North Carolina who began teaching in 1975. When they compared data across districts that had different salary schedules, they concluded that a $1,000 increase at each salary step increased the teachers' duration in that district by two to three years.[14] In addition to offering financial incentives, we might follow the lead of European and Asian countries and provide them opportunities for better job preparation and professional development.[15]

Data from the National Center for Education Statistics and the National Educational Goals Panel show that Connecticut has been highly successful in K–12 schooling, even though poverty rates in the state have been increasing. Connecticut fourth graders and eighth graders placed first in the nation in National Assessment of Educational Progress (NAEP) testing and the fourth graders placed first in the nation in mathematics.[16] How was the state able to achieve these results? By using "teaching standards, followed later by student standards, to guide investments in school finance equalization, teacher salary increases tied to higher standards for teacher education and licensing, curriculum and assessment reforms, and a teacher support and assessment system that strengthened professional development."[17] Part of Connecticut's success was a strategy to raise teachers' salaries. During a five-year period the average teacher's salary in Connecticut was increased by just over $18,000 a year.

The stunning educational reforms in Connecticut were reflected in SAT scores. In 2004, 85 percent of Connecticut high school graduates took the SAT, compared with a national average of 48 percent.[18]

Linda Darling-Hammond, the nation's leading researcher about teachers and teaching, argues that while testing isn't the answer, it can provide information that leads to better accountability:

> More successful outcomes have been secured in states and districts . . . that have focused on broader notions of accountability, including investments in teacher knowledge and skill, organization of schools to support teacher and student learning, and systems of assessment that drive curriculum reform and teaching improvements.
>
> . . . there is evidence that high-stakes tests that reward or sanction schools based on average student scores can create incentives for pushing low-scorers into special education, holding them back in the grades, and encouraging them to drop out so that schools' average scores will look better.[19]

Improving schooling requires that we recruit more quality teachers and that we work to keep them in the profession. This means significantly increasing the respect we give teachers, the financial compensation we give them, and the professional development we provide.

Teachers and the Achievement Gap

Teachers and mentors are critical to closing the achievement gap. In the past few years, CGU professor Gail Thompson has published a series of books about how to improve the education of African American students and other students of color. She observes: "When it comes to African American and Latino students, in my opinion, too many teachers waste time focusing on their perceived deficits, which often are merely cultural differences that teachers measure against middle-class norms. One of the recommendations I always make to teachers who want to increase their efficacy with students of color is to stay away from the negative teachers on campus."[20] She then presents specific strategies that such teachers can use: "Let students know you care; share the real you, by letting them see you are a real human being; have high expectations; make the classroom relevant to the real world; showcase their talent; encourage them to synthesize; assign regular, beneficial homework; and encourage students to write letters to authors of books they have read."[21]

Engaging Students

Effective teaching is not achieved through the use of a mechanical model. Learning involves the active engagement of the learner. He or she must incorporate the new material, linking it to his or her existing knowledge, based on both experience and prior learning. These considerations make it all the more important that teachers at both the precollege and college levels find ways to connect to the frame of reference in the student's mind. In my experience, the method that seems most promising is the frequent use of metaphors or stories. Regardless of the method they feel works best, teachers must know how to present the same idea in different ways to respond to the changing needs of their students.

ADDING VARIETY TO TEACHING

When I was a child, I studied magic. My hero was John Scarne, who was perhaps the greatest card magician of all time. He once said about card tricks, "If you know 100 ways to determine which card a person picked and one way to present it, it looks as though you know one trick. If you know one way to find out which card a person picked, and 100 ways to present it, it looks like you know 100 tricks."

When I was 19, I lived in a suburb of Honolulu called Aina Haina. I worked in the forklift shop of the Dole Hawaiian Pineapple Company driving a small tractor. My job was to pick up freshly canned pineapple products and deliver them to the machines where the labels were put on the cans. The pineapple cans were placed on trays, the trays were stacked on carts, and 10 or 12 carts would be linked together in a train. I linked my tractor to the first cart and pulled the train through the cannery in and out of warehouses. On the return trip, I pulled a train of empty carts back to the point of origin.

Watching the labeling machines, I learned a valuable lesson in commerce. In point of fact, I was relearning the lesson I had learned from John Scarne. Canned pineapple from the same batch might be labeled as Dole Pineapple, or as the Grand Union brand (a popular supermarket chain at the time), or as S&W Pineapple (a "high-end" gourmet product line).

One skill that is indispensible for an effective teacher is knowing how to present the same idea in many ways. A true educator does not present a concept—say, the standard deviation and its calculation—once and then dismiss as inadequate those students who don't grasp the concept immediately. A true educator will present the same concept in several ways, until one of those presentations clicks with a given student. It is important to remember that each student brings a unique frame of reference and catalog of experiences to the classroom. It is a teacher's job to find out how to connect to that student's frame of references.

The instructor must learn about the students' frame of reference to improve the quality and effectiveness of his or her teaching. The benefits of doing this are reciprocal.

Consider this reflection by Parker Palmer, a widely respected scholar who has written extensively about the teaching process:

> Good teachers possess a capacity for connectedness. They are able to weave a complex web of connections among themselves, their subjects, and their students so that students can learn to weave a world for themselves. The methods used by these weavers vary widely: lectures, Socratic dialogues, laboratory experiments, collaborative problem solving, creative chaos. The connections made by good teachers are held not in their methods but in their hearts—meaning *heart* in its ancient sense, as the place where intellect and emotion and spirit and will converge in the human self.[22]

Or this reflection by award-winning English teacher Edward Blanchard:

> The most important focus of any instructional scheme is the relationship between the teacher and the taught. These appellations unfortunately suggest a linear transfer of discrete packages of knowledge from the possessor to those in need. Language needs to be more precise in describing this complex process. We need to introduce the mystery of human relationships and art here. Gaining trust and inspiring people is an art. It is not psychological manipulation or magic, but an interaction which is dependent upon authenticity and love above all else. Success results from realistic, down-to-earth communication, and love of subject combined with mutual respect and a dispositional attitude which lends itself to an atmosphere of positive discussion.[23]

Scientific research, theorizing, and model-building are enriched, and may be transformed, if we learn more about the cultural perspectives underrepresented groups bring to the table.

I believe that teachers must approach students with a new philosophy. First, their fundamental goal must be for the students to learn as much as possible. This is the bottom line, and it should not be eroded by other, conflicting teacher motivations, such as demonstrating intelligence, garnering a paycheck, or scaring students by emphasizing how tough the discipline is. Second, teachers must believe that all students can master the material. The question is not *whether* the students will learn but *how* they will learn.

Teachers must teach to the students' frame of reference and tie new material to their lives. Students and teachers should be engaged in the construction of meaning, not the transmission of knowledge. For example, in every introductory statistics course, I must teach about variance and standard deviation, concepts that most students do not find intrinsically exciting. The notion is that while two samples may have the same mean or average, the variation or dispersion around those means may differ. Consider the ages of six children at two child care centers, Center A (4, 4, 4, 4, 4, 4) and Center B (2, 2, 2, 6, 6, 6). They exhibit the same mean, but any teacher or parent can tell you that there is a great difference between working with six 4-year-olds and working with three 2-year-olds and three 6-year-olds.

In my teaching, I illustrate the construct of variance and standard deviation by citing a powerful personal experience of Stephen Jay Gould. He emerged from exploratory surgery in the mid-1980s to be told by his physician that the surgery had revealed a particularly virulent form of cancer and that the median life expectancy for patients with this cancer was six months. Since the median is a measure of central tendency, Gould needed to learn about the variation and the shape of the distribution. He subsequently discovered that even though half the patients with this form of cancer died within six months (the *median* life expectancy), many patients survived for many years. Gould read the medical literature and discovered that he had many of the characteristics of patients who survived a long time.

The use of stories to communicate concepts should not be confused with presenting anecdotes in place of quality teaching. Once, after giving a lecture to 200 UCLA undergraduates, I described to a colleague the range of student behaviors I had observed. Some students had listened attentively; others had slept, snapped their chewing gum, sipped coffee, read a newspaper, joked, or fondled one another. "You don't understand," my colleague replied. "This is the television generation. They don't realize that you can see them!" And they want to be entertained.

Effective teachers lead students to master and enjoy skills, techniques, and knowledge that they thought they could never acquire. The challenge is to extract and nurture the talent and interest that the student already possesses. This approach is consistent with the epistemology and pedagogy of the late Brazilian educator Paulo Freire. According to Marilyn Frankenstein, "Freire is adamant that the content of an education for critical consciousness must be developed by searching with the students for the ideas and experiences which give meaning to their lives."[24]

Returning to statistics, a course I teach: this is a subject that is required in many graduate and undergraduate programs and occasionally is offered in high school. Most students dread statistics and put off taking it as long as possible. Furthermore, few students remember the material six months after the final exam, and almost none could effectively employ statistical analysis in research or in personal decision making.

The reasons for student aversion to statistics lie not in the odious nature of the subject (some of us truly enjoy it) but in how it is presented—and by whom. Statistics (and mathematics and science more generally) should be taught by linking the subject to the student's experience base and illustrating its application in real research. Unfortunately, people who master and teach statistics do not always possess superior interpersonal skills. There is an old joke that goes like this: Question—How can you tell if a mathematician is an extrovert? Answer—He looks at *your* shoes when he is talking to you.

Here is the way statistics is often taught and what is wrong with these approaches:

- Professors stand at the board deriving equations. Derivations teach students little that is useful; what students need to know is when and how to apply each technique.
- Students are expected to memorize equations and spit them back in closed-book exams. Researchers don't need to memorize equations; they can always look them up. All statistics exams should be open-book, a structure that more closely mirrors how research is done in reality.
- Statistics courses often have mathematical prerequisites like calculus. While such prerequisites serve the useful function of scaring away some students and easing the professor's workload, imposing these restrictions

has no pedagogical basis. Students can master applied statistics through multivariate analysis even if they know only a little algebra.

- Students are taught and dutifully master an array of techniques—chi square, t test, correlation, regression—but they are not shown why or when to choose one technique over another.
- Too many professors simply show students how to employ statistical programs (the most widely used being the Statistical Package for the Social Sciences, or SPSS) to grind out statistics and skip the important material—namely, what the technique is and how to use it.

Each of these approaches drives students from statistics for the same reason that many junior high school students are turned off by science: the subject matter isn't linked to students' past experiences or future work.

Students must be shown how research is conducted and how statistical methods are used. From this perspective, some of the material in standard texts is useless. Many statistics textbooks devote about half a chapter to the index of qualitative variation. Yet, in more than 40 years of doing research, observing my colleagues' research, and reading studies, I have yet to encounter an index of qualitative variation. Why burden students with something they will probably never need?

Stories and other illustrations can link abstract mathematical techniques to the student's frame of reference. For example, I have found it helpful when introducing the concepts of probability, random fluctuations, and statistical significance to describe some highly improbable experiences from my own life. (Once, in trying to salvage a magic trick that had floundered, I handed the Washington, DC, phone book to a friend. I instructed her to pick a name at random and said that I would tell her that person's phone number. I was bluffing, of course. But she closed her eyes, pointed, and read me the name of my brother-in-law! The probability that she would pick at random someone whose phone number I knew from that huge phone book was infinitesimal.)

In *The Call of Stories*, child psychiatrist and teacher Robert Coles describes the power of stories to instruct, to connect to people's lives, and to communicate. In the following passage he describes how the concept "stories" enabled him to make a breakthrough in communication with a recalcitrant psychiatric patient.

> She was still hesitant to begin, so I said it—said what no one had suggested I say . . . : "Why don't you just tell me a story or two?"

> . . . She looked at me as if I'd taken leave of my senses. I began to think I had: this was no way to put the request I had in mind. Why *had* I phrased my suggestion that way? I explained that we all had accumulated stories in our lives, that each of us had a history of such stories, that no one's stories are quite like anyone else's, and that we could, after a fashion, become our own appreciative and comprehending critics by learning to pull together the various incidents in our lives in such a way that they do, in fact, become an old-fashioned story.
>
> . . . For the first time in my short career in psychiatry I saw a noticeable and somewhat dramatic change take place in a patient—and not in response to any interpretation or clarification of mine, but merely as a result of a procedural suggestion, as it were: how we might get on, the patient and I.[25]

A Master Teacher

Consider the aptly named Dave Master. Master is a highly effective teacher, who, as an art teacher in a small community, transformed the lives of countless students and who, in the process, transformed the motion picture industry.

About 30 years ago, Master was hired to teach art in a junior high school in Rowland Heights, California, a blue-collar, multiethnic community 25 miles east of Los Angeles. On the surface, there was no reason to expect any of the students in that community to excel academically. Many got into trouble and many left school before graduation.

A totally committed teacher, Master had the novel idea to teach his students about animation. He scrounged his own equipment. He worked long hours with the students. Some students undoubtedly would have dropped out of school were it not for the love of animation Master's teaching engendered. Eventually he moved to the high school in Rowland Heights and continued teaching animation. From the beginning, he connected with international experts, the "grand old men" of animation, and invited them to visit his school and to instruct and advise his students. This began when he attended a lecture by Stephen Leiva, the publicist for legendary animator Chuck Jones and patiently waited to talk with Leiva, who later introduced him to Jones: "Jones reviewed the students' primitive early work and, in his endearing and straightforward way, said, 'I love your enthusiasm, but you don't know what the hell you are doing!' . . . Jones took Master under his wing."[26]

The results were amazing. Some of Master's former students, now in their 30s and 40s, hold key positions in the growing field of film and television animation. Most professional animators in the industry were educated in one of five institutions. Four are colleges in the United States (two), the United Kingdom, and France. The fifth is Rowland Heights High School. Several of Master's former students worked on the blockbuster film *Avatar.* One of his students, Musa Mustafa, was appointed to the Atari advisory board at the age of 15. Debra Feinstein notes that "many of his students went directly from high school to entry-level positions in the [animation] industry" and goes on to describe why this was the case:

> Master . . . rewrote the rules of how to educate young artists. Rowland felt more like a Hollywood studio than a classroom. Students had access to $500,000 worth of computers and video equipment. They didn't just worry about report cards; they created portfolios of their work. They didn't just take classes; they made short films and competed for awards.
>
> "We had them doing physics, calculus, animation, sculpting, and working with computers, all in the same setting," Master says. "They learned management skills, how to have a vision, how to pitch their ideas and get other people behind them, how to manage equipment, schedules, and budgets. They learned things as they needed to know them, which is the best way to learn."[27]

Master notes that two of his students

> used animation as a context to explore their technical, engineering, and mathematics interests. In a classroom that honored and appreciated both pathways, these students were able to not only gain expertise in their area of interest, but also develop skills and abilities in cross-disciplinary areas. After leaving school, many of the jobs they pursued *required* cross-disciplinary sensibilities. My students were ahead of the game because of their trans-disciplinary experiences in my classroom. In the 21st century this kind of educational experience will become essential. "Silo" type instruction prepares students for a world that is disappearing.[28]

Dave Master has won numerous awards for his dedicated teaching in this little-known high school in a little-known town. He has been asked to lecture at Harvard. He was selected as the IBM / Technology and Learning National Teacher of the Year. Now he has leveraged his impact by teaching online. From a studio in Burbank I watched him instruct students at an Alabama high school.

Several years ago, Master invited me to a party he held for his former students who now worked in the field of animation. I met one man, now in his late 30s, who had been getting into serious trouble as a teenager and was on the verge of dropping out of school when he entered Dave's art class. Today he is prosperous and successful, and he has won three Emmys.

I spoke with John Ramirez, now a highly successful animator and theme park designer.[29] He has applied his creative talents to theme parks in Singapore, Korea, and Japan, as well as in the United States. He calls Master's class "unlike any other." Master's class had a structure, but it was invisible to the students. "He had set up the goal posts and we didn't even know it," said Ramirez. He said Master acknowledged that he was learning as he went, constantly changing directions.

Mr. Ramirez estimates that more than 1,000 of Master's former students now are working in the industry. "Without his class, I don't think the animation industry would be what it is today." As students, "we felt real ownership. It was our program. If I walked into a math class, it was my teacher's class." John talked about the visits by Chuck Jones, Frank Thomas, and Ollie Johnson, all legendary animators. "It was like a conservatory." Master "could have taken any subject and made it engaging. He taught us how to think."

One of those students, Katherine Concepcion, has been a producer at Animal Logic, Sony Imageworks, and Warner Bros. Another, Jennifer Cardon-Klein, won many awards in local and national film competitions even as a student. She worked on Warner Bros.' first animation feature film. Her husband, Bert, also a graduate of the Dave Master program, has worked on 10 Disney films.

Mike Belzer is another former student. He first studied animation at the age of 12. He has worked at Disney, Warner Bros., Pixar, and, presently, Valve, "the coolest game studio on the planet." His work on feature films, TV, and video games includes *Bolt, Meet the Robinsons, Kangaroo Jack, Dinosaur,* and *Gumby,* as well as the advertising icons Hershey's Kisses and the Pillsbury Doughboy. He says, "Having seen many teachers, Dave is unique. I've never seen anyone who comes close." Dave was "the great ringmaster, but he never let his ego get in the way of accomplishing his goals." Sometimes his approach was, "I don't know, but let's find out together."[30]

Now consider this observation by Jay Parini, a distinguished novelist, poet, and professor, who has written about the teaching process:

> My notion of the ideal teacher is that of *primus inter pares,* with the teacher as lead student. I wish I had understood from the beginning that I was, at heart, a perpetual student: amazed before the world's variety and unworded beauty and frustratingly contradictory nature. As student and teacher in one skin, I

> work at unraveling the many strands of this world, putting into words its silent beauty, and attempting to resolve the contradictions. Success, in these terms, is always a kind of failure as well, and demands a fresh start, a willingness to ask the fundamental questions in an innocent way, a need to set the whole dialectic in motion once again.[31]

Belzer adds that Dave Master resisted administration pressures to assess student progress with standardized tests. Belzer praised Master's courage and efficacy. "He would have had the same impact if he taught something other than animation." Master "changed my life. It was not what he taught me, but how he taught me. He let me learn how to learn." Master also changed the lives of students who did not go into animation. It was more than an animation class; for example, they learned about the management of people.

As a student, Marci Gray was drawn to the management and human relations components of the Rowland animation projects. She became a secretary and administrative assistant to Master and later followed him to Warner Bros. She is presently manager of human resources at Warner Bros.—in other words, the artistic recruiter. She said, "Dave figured out what your strengths were and then placed you in situations where you would exercise that skill set. Not everyone is good at walking the high wire or juggling." She adds, "We loved being in the program because of the energy he created. He always led with a vision. . . . He pushed you to your limits. He set higher expectations for people than they did themselves and most of the time they would hit them."[32]

We should not simply seek and rely upon charismatic teachers like Dave, but rather should try to bottle what they do. I asked Dave what principles guided his teaching. He answered:

> What you teach is not as important as how you teach it. The key is not to teach isolated techniques, but, from the start, to embed student learning in real life projects. Through these projects, my students applied the full range of fine arts standards (i.e., making films from storyboarding, animating, and layout design through editing, to designing full sized "open house" environments and animatronic devices, with full project management and budgeting throughout the process).
>
> The students find this more relevant and engaging, less contrived and abstract. They became part of a field; this built a bridge to industry.
>
> Connection to, communication with, and critique from, professionals and experts in both the field and academia was critical to my students' development and learning and for *my* professional growth. I saw myself as a curriculum

> designer and then orchestrated periodic opportunities for my students and me to have experts in the fields critique our efforts and provide us with advice (I called this the "Who sez?"). Without this central activity, our successes would not have been possible.
>
> I made sure that I selected professional role models that were from both genders and from diverse racial, cultural and ethnic backgrounds. These models embodied more than professional expertise, they gave my students the confidence that they could break through real and imagined barriers.

He concluded, "In most schools we are preparing students for a world that does not exist post-graduation."[33]

Not surprisingly, Master attributes his success to his own teachers: "These great teachers taught me how to connect my students' hearts, heads, and hands to their futures. I was lucky to have a few teachers that made a difference in my life. They didn't let the outside world crush their enthusiasm. They didn't let infighting drain their energy. My entire life is indebted to them. For well or ill, every teacher touches countless lives. The teaching profession is the most important job in the world. Teachers are responsible for nurturing the future."[34]

Technology and Teaching

As it was in Dave Master's classroom, technology is often crucial to delivering quality education. But like any new innovation, teaching with technology invites controversy. In debates about the effective use of computing technology in the classroom, critics make three arguments:

1. The teacher must use a prepackaged curriculum.
2. The more students work on computers, the less they will interact with teachers.
3. The latest hardware and software are more available to affluent students.

Each of these arguments has some merit. But in an earlier era, similar arguments were made to oppose the introduction of books into the classroom.

Unfortunately, schools and school districts that do embrace technological innovations often do not implement them very well. They may spend most of their technology budgets on hardware and software and spend relatively little on training, which averages no more than 15 percent of all technology funds.[35] Teachers report limited time to practice technology skills, to develop new activities, and to conduct activities during a tightly scheduled school day. Teacher limitations were reportedly the most significant barriers to increased integration of educational technology into the classroom.[36]

Often education leaders act like some perplexed British officers during World War II, who were concerned about the seemingly wasted motions used by artillery personnel firing cannons left over from World War I. Junior officers were at a loss to explain these moves, which had no discernable purpose. Finally, a retired general saw a film of this behavior, snapped his fingers and said, "It's easy. They're reining in the horses." Faced with a bewildering and constantly changing array of reforms, far too many teachers and administrators are "holding the horses." They need guidance in breaking old habits, including a crippling reliance on jargon. Old habits die slowly. There is a saying in the Navy that if you dig deep enough in a modern battleship, you will find sails.

In terms of teaching, our use of technology always should be informed by what social scientists and other observers teach us about the history of successful innovation.

A creative reporter for the *Wall Street Journal* once began a 1995 column this way: "A new century is at hand and a fast-spreading technology promises to change society forever. It will let people live and work wherever they please, create dynamic new communities linked by electronics, improve the lot of the poor and reinvent government—unless its use for illicit purposes sparks a crackdown."[37] The reporter noted that such predictions were made at the beginning of the 21st century about the Internet and also at the beginning of the 20th century about the telephone. He added: "Telephones were blamed for insanity and for a total loss of privacy. . . . Sociologist Charles Horton Cooley said, 'an intricate mesh of wider contacts' would turn people's geographical neighbors into strangers. . . . It didn't happen."[38] Unfortunately, this insightful social observer and reporter became better known for the nature of his death than for his work. He was Daniel Pearl, who was killed by terrorists shortly after 9/11.

Persuasive or Empowering Technology

Within the field of information technology, the use of new devices to help people change attitudes and behavior is called "persuasive technology." This discipline was created by B. J. Fogg, an information scientist at Stanford University. According to Donald Norman, one of the nation's leading experts on human/computer interactions: "Today's technology is used to change attitudes and behavior. This creates powerful opportunities, multiple challenges, and severe ethical issues."[39] I prefer to think of these new devices and applications as "empowering technologies."[40] Dr. Fogg cites six advantages computers have over traditional media and human persuaders. They can

1. Be more persistent than human beings
2. Offer greater anonymity
3. Manage huge volumes of data
4. Use many modalities to influence
5. Scale easily
6. Go where humans cannot go or may not be welcome[41]

A familiar example of persuasive technology cited by Fogg are the suggestions made by Amazon.com for other books a customer might also want to purchase. Amazon's software allows the site to link books a customer has purchased to similar books he or she may not have seen.

My colleague Samir Chatterjee and I have been exploring the ways in which the latest technology can improve education. Our focus has been health education. Obesity is a major health problem, if not *the* major health problem, in the United States. And we have conducted preliminary research about using technology in obesity reduction programs. You cannot have a productive, high-tech workforce if workers are not healthy.

Chatterjee and I developed proposals, received external support, and conducted preliminary research about the application of new technologies to reduce obesity rates, prevent stroke, and treat post-traumatic stress disorder. We wanted to find out how we can help people develop diet and exercise habits to lose weight. To do so, we enlisted the participation of volunteers willing to submit various physiological measurements for study. Our strategy involved combining GPS technology, cell phone communication, and special devices using wireless technology to measure vital signs like blood pressure.

The proposed system works something like this: A participant might receive a text message that says, "We noticed that you have just entered a Pizza Hut. We certainly hope that you will order a salad." Our research is guided by social psychological theory about attitude change and about behavior change. Simply put, we hypothesize that obese people will be more successful in losing weight if they have the support of professionals and other program participants and if they are willing to undergo constant monitoring during the weight reduction program.

Similar technologies can be used to assist students who are struggling with mathematics and science homework. Technology has the potential to facilitate and accelerate student mastery of STEM fields. But technology also can undermine the development of analytical reasoning and deeper learning. Simply using the latest hardware and software for instruction will not solve all our problems.

Conclusion: *We don't want to miss out on the benefits of new technology, especially computer-based technology. We must, however, be alert to the potential negative effects of new technology and we must keep our fundamental educational goals constantly in mind.*

Technology's Limits

Let's face it: Most people, from elementary school students to senior research scholars, now use Google to find information. But students rely too heavily on Google at their peril, and their analytical reasoning may be impaired by this dependency.

Memory researchers talk about the difference between recognition and recall. Suppose I present you with this multiple choice question:

> The United States fought against Italy in which war?
> a. The War of 1812
> b. World War I
> c. World War II
> d. The Gulf War

The correct answer, of course, is c. Most people will be able to identify that answer. That is called recognition. But what if I asked an open-ended question instead:

> In which war did the United States fight against Italy?

I would wager that a smaller percentage would be able to produce the correct answer. When today's students enter the workforce, they will be challenged to retrieve knowledge and to apply it. The cognitive skills required will go way beyond recognition skills. The cognitive skills required are much more likely to be developed in a curriculum that emphasizes analytical reasoning and deemphasizes computer-based searches, than in a curriculum that relies heavily on Internet search engines. In his book *A Whole New Mind*, Daniel Pink argues that the most important people during the next century will be those who think with the right side of their brain (which is commonly thought to be the creative side), not those who have developed only the left side of their brain and are technologically skilled.[42]

A LONE INNOVATOR

Jerry Lucas is a determined, talented, disciplined, and goal-oriented man. He has had two successful careers and now designs educational software as his third career.

As a boy, Lucas wanted to be a successful basketball player. He practiced diligently, setting a goal of 5,000 shots at the hoop every day. He wasn't satisfied just to make the shot; he had to sink it from various locations on the court, picking the part of the rim where the ball would hit the net.*

Lucas became one of the greatest basketball players ever. He was the first player in history to win championships at four levels: a high school state championship, a college national championship, an Olympic gold medal, and a professional world championship. Only two other players have accomplished that since. Lucas was twice named national high school player of the year and twice national college player of the year. He was all-pro seven times. At the age of 21, he was featured on the cover of *Sports Illustrated* as "Sportsman of the Year," the best athlete in the world.†

Virtually every pro athlete finds it difficult to retire. Scott Tinley, a Hall of Fame triathlete, who either won or placed second in the Hawaii Ironman competition six times, has written a book about the struggles and adjustments he, and other successful athletes he knows, faced. He quotes legendary quarterback Archie Manning: "I miss football so much—heck, I even miss the interceptions."‡

Jerry Lucas had no difficulty walking away from sports. He always had had an active mind. As a child, he was often bored and restless, and he loved to play mental games: "I remember looking at an oil company billboard and saying to myself, 'What would "SHELL" look like if the letters were arranged in alphabetical order?' I mentally rearranged it to "EHLLS" and I was hooked. Ever since then, I have memorized words alphabetically as well as normally."§

He learned how to memorize all kinds of information. He insisted on attending Ohio State University on an academic scholarship, not an athletic scholarship, and he graduated Phi Beta Kappa. After retiring from sports, he focused on developing his memory system and teaching it to others. Lucas published a number of books about memory; one of them, co-authored with Harry Lorayne, has sold more than 2 million copies.

*Jerry Crowe, "Innovation Comes Naturally to Basketball Hall of Famer Jerry Lucas," *Los Angeles Times*, Jan. 10, 2010.

†www.jerrylucas.com.

‡Scott Tinley, *Racing the Sunset* (Guilford, CT: Lyons Press, 2003), p. 201.

§Harry Lorayne and Jerry Lucas, *The Memory Book: The Classic Guide to Improving Your Memory at Work, at School, and at Play* (New York: Ballantine/Random House, 1974), p. xi.

Recently, Lucas designed a Web site that will embed his memory-learning system in a series of fun games and challenges. When students have completed this series, they can enter a virtual world with many planets, "Dr. M's Universe." On one planet they can play a popular video game. On another they can play a game that will teach them mathematics. Another planet and game will be about history, another about grammar. Other planets will teach youngsters about diet and exercise.

He is determined to make this virtual world fun, in part by creating it with the best available technology. The images will be Disney/Pixar quality and three-dimensional, developed by an artist who was trained by one of the "nine old men" of animation and who has worked with Steven Spielberg and George Lucas. Lucas has finished his business plan and has a promotional DVD. He is seeking venture capital and has no doubt he easily will raise the resources he needs.

This virtual world will be a profit-making venture. But it will not be expensive. The monthly fee will be less than $15, about what a young person would pay for access to only one of the popular video games.

Mr. Lucas considers this project the culmination of his life's work. He says, "My goal is to revolutionize the educational system in America to make it fun, easy, and long lasting for all involved."*

I wouldn't bet against him.

*Personal communication, Jan. 26, 2010.

Students today engage in many electronic activities simultaneously. I doubt that multitasking students perform a given task as effectively as someone who focuses only on that task, but the thrust of today's technology is toward multitasking. A teenager today might simultaneously listen to music on her iPod, e-mail friends, send instant messages, update her Facebook page, text from her cell phone, and chew gum. After days, months, years of being stimulated electronically this way, will she be able to sit quietly and focus on a mathematics problem, or on any other problem? Some have labeled this trend "the attention-deficit society."

In Conclusion

The research reported in this chapter clearly indicates that the key strategy for improving STEM education in America is to improve teaching. In part, this means recruiting and retaining creative and committed teachers through programs like Math for America training and Noyce scholarships. It also requires innovative preservice education of these teachers, as in the CGU Teacher Education Internship Program. And it means revitalizing the knowledge and skills of current teachers through professional development. The focus should be less on who teachers are and more on what they do. Ideally, teachers would perform optimally and derive great meaning from their work, a condition that is described by Mihalyi Csikszentmihalyi:

> We have seen how people describe the common characteristics of optimal experience: a sense that one's skills are adequate to cope with the challenges at hand, in a goal-directed, rule-bound action system that provides clear clues as to how well one is performing. Concentration is so intense that there is no attention left over to think about anything irrelevant, or to worry about problems. Self-consciousness disappears, and the sense of time becomes distorted. An activity that produces such experiences is so gratifying that people are willing to do it for its own sake, with little concern for what they will get out of it, even when it is difficult, or dangerous.[43]

I close with the recommendations about teachers presented by the task force that produced *Rising above the Gathering Storm*:

> Recommendation A: Increase America's talent pool by vastly improving K–12 science and mathematics education.
>
> - Action A-1: Annually recruit 10,000 science and mathematics teachers by awarding four-year scholarships and thereby educating 10 million minds. Attract 10,000 of America's brightest students to the teaching profession every year, each of whom can have an impact on 1,000 students over the course of their careers.
> - Action A-2: Strengthen the skills of 250,000 teachers through training and education programs at summer institutes, in Masters' programs, and in advance placement (AP), and International Baccalaureate (IB) training programs.[44]

CHAPTER 5

Mentors and High Expectations

> The carver holds the unworked ivory lightly in his hand. Turning it this way and that, he whispers, "Who are you? Who hides there?" And then, "Ah, Seal!" He rarely sets out, at least consciously, to carve, say, a seal, but picks up the ivory, examines it to find its hidden form and, if that's not immediately apparent, carves aimlessly until he sees it, humming and chanting as he works. Then he brings it out: seal, hidden, emerges. It was always there: he didn't create it; he releases it; he helped it step forth.
>
> Carpenter, Varley, and Flaherty, *Eskimo*

Once a student decides to attend college, we want him or her to succeed, to achieve, and to graduate. The attrition rate in America's colleges is a national disgrace. The loss of talent represented by the attrition of STEM majors is significant and disturbing. Far too many STEM students either transfer out of the sciences or drop out of college. This problem is more severe for women, students of color, and students from poverty. The achievement gap, first evident in elementary education, persists through undergraduate education.

But we now have learned how to confront and close that achievement gap.

The U.S. Department of Education examined the persistence of college students in STEM disciplines, drawing upon data from the Beginning Postsecondary Students Longitudinal Study, covering the period from 1995 to 2001.[1] The researchers in this study also drew upon data from the 2003–2004 National Postsecondary Student Aid Study and from the Education Longitudinal Study of 2002–2006. They reported that in 2003–2004 about 13.7 percent of undergraduates were in a STEM field, broken down as follows:[2]

computer/information sciences	4.9%
engineering	4.2%
biological/agricultural sciences	3.1%

physical sciences	0.7%
mathematics	0.5%

These researchers also examined degree attainment and persistence in STEM fields as of 2001. They found that, of all students who had entered a STEM field in 1995–1996,

- 37.1 percent had attained a degree or certificate in that field,
- 7.5 percent were still enrolled in a STEM field,
- 27.1 percent had transferred to a non-STEM field, and
- 28.3 percent had dropped out of college without a degree or certificate.

In this chapter, I suggest that attrition and the achievement gap are directly connected to the reward structure for faculty members and the low value placed on teaching and mentoring in many colleges and universities. I argue that the most important solution to the attrition problem is the presence of faculty mentors who hold high expectations for their students.

Confronting a Culture of Exclusion

Sheila Tobias devised an intriguing ethnographic experiment to discover why great numbers of very capable college students avoided science. She hired seven postgraduates from a variety of fields in the humanities and social sciences, each of whom was high-achieving and literate. All but one had avoided science totally in college, and none had selected a scientific career. These researchers were asked to audit seriously a semester-long, calculus-based course in physics or introductory chemistry. They were asked to focus on why students like themselves might find introductory science "hard" or even "alienating." Each student was asked to keep extensive notes on his or her experiences and observations about the course, the instructor, and the material and to write a final essay drawing together those observations.[3]

One of these postgraduates was "Eric," a summa cum laude graduate in literature from Berkeley. In his notes he commented on the value of cooperative study groups, which are discussed later in this chapter: "My class is full of intellectual warriors who will someday hold jobs in technologically-based companies where they will be assigned to teams or groups in order to collectively work on projects. [But] these people will have had no training in working collectively. In fact, their experience will have taught them to fear cooperation, and that another person's intellectual achievement will be detrimental to their own." Tobias notes that Eric thought that one consequence of students' doing their work "in private" was "the absence of any opportunity for them to talk about the physics they were studying. They seemed inhibited, he observed, even about asking questions."[4]

Eric concluded that while the material covered in a physics course like the one he was observing was inherently difficult, the way it was taught made it much more difficult and, for some students, impossible to master. He noted that the instructor raced through the material, giving students little time to digest one topic before tackling another. Students were reluctant to speak out, to ask questions, and to articulate what they did or did not understand. These problems were exacerbated by the competitive nature of the grading process and by the significance this grade held for the students' academic and career development.

According to Walter Massey, former director of the National Science Foundation, "The common concept of 'success in science' . . . seems to have created an illusion that only 'the best and the brightest' can do science. Course work is viewed by many faculty as a way to separate the 'men' from the 'boys.' Unfortunately, these courses also tend to separate the men from the women—and the white men from just about everyone else."[5]

Confessions of a Nonscientist

In his article "Confessions of a Nonscientist," Bill Long, a Presbyterian minister, notes that it was assumed that he would become a scientist. He won science awards in high school. His father was a pioneering computer designer, and his older brother was studying science at Rensselaer Polytechnic Institute. He had received top SAT scores in mathematics and science. However, once in college, he quickly transferred out of the sciences. He found that his science and math professors "seemed terribly indifferent, or even hostile, to me, the student." It was if they were saying to him:"You students are really worth very little. Most of you will not be around very long, so I'm not going to be too concerned about you. If you sink, that is fine with me. If you swim, I will perhaps notice you, but even in that case, I will probably not get around to you for a few more years. I have so many more important things to do than to be with you."[6]

The problems in college teaching partially explain the trends and changes in students' decisions to major in science, but these problems do not occur in a vacuum. They are related to how prestige is awarded in academia, which, in turn, is linked to funding patterns for university research, which powerfully influence how professors spend their time. We must break this destructive cycle and replace it with a more engaged and committed approach to teaching. But does that mean that American professors and universities must cease to be world leaders in research?

Publish or Perish

John Slaughter, former president of Occidental College, put it this way: "Research is to teaching as sin is to confession. Unless you participate in the former, you have very little to say in the latter!"[7] Some have argued that a professor can be an excellent teacher or an excellent researcher but not both: each role is demanding, and there simply is not enough time in the day to do both well. Others, including myself, believe that it is possible to excel at both tasks. There are many examples of world-class researchers whose teaching is superior.

Kenneth Feldman conducted a meta-analysis of studies about the links between research and teaching.[8] He estimated the mean correlation reported in these studies between research and teaching excellence to be .13. To put it differently, this means the percentage of one variable associated with the other is less than 2 percent. In short, *there is almost no relationship between research and teaching excellence. Good researchers are neither necessarily good teachers nor necessarily bad teachers. Universities should be hiring outstanding researchers who are also outstanding teachers.*

The faculty members at a university can easily discover what their administration really values. Confusing rhetoric can always be clarified by examining who gets promoted or by comparing teaching loads across institutions. They vary incredibly. It is revealing that administrators recruiting bright young PhDs refer to "research opportunities" and "teaching responsibilities." The latter are often measured in terms of "contact hours," as though one were being exposed to an infectious disease.

My colleague Paul Gray and I have published a small book of candid advice to new faculty members.[9] This book is not a scientific study. Rather, it is a compilation of our experiences, observations, and insights about how the academic world really works. It is what we would say to a new faculty member if he or she invited us to lunch and asked for advice about how to achieve tenure. Here are some relevant excerpts.

1. Gray's Theorem of N + 2. The number of papers required for tenure is N + 2, where N is the number you published.

   Corollary: Gray's Theorem is independent of N.

4. Drew's Law. Every paper can be published somewhere. Your first papers will be rejected. Don't worry about this. View the reviewer's complete misunderstanding of your brilliance as cheap editorial help. Use his or her advice to revise. Every paper has a market. If *Journal A* rejects it, make the appropri-

ate changes and send it to *Journal B*. If the work is sound, someone will publish it.

5. Make sure you have a mentor early in your career. The old apprentice system still exists. Try to find mentors who were successful with others, who will support you, and who believe that furthering your career helps their own career. Such a mentor is preferable to the internationally famous Nobel Prize winner who exploits you.

39. Teaching is a great personal satisfaction and an important public good that you perform. However, publications are your only form of portable wealth.

62. Tenure committees look almost exclusively at publications that appear in peer-reviewed journals or in scholarly books. It is, in a sense, a tragedy that you get much more credit for what appears in a "write-only" journal (i.e., a journal with minute circulation) than for what appears in a high-circulation, widely read popular magazine. But that is the way the game is played.

These were our observations about how the system presently works. We need to change that reward structure. We can be guided in that transition by gifted teachers and mentors.

Exemplar Mentors

> That, of course, is the essence of teaching—taking chances. And you can only do that if you are willing to come down from your perch as a professor of this or that and be as vulnerable (or almost as vulnerable) as your students. No professorial vulnerability, no real teaching.
>
> Page Smith, *Killing the Spirit*[10]

Jeanne Nakamura and David J. Shernoff studied three outstanding geneticists and mentors, focusing on their "lineage" (that is, who they mentored and who those former students mentored). They identified six key elements of successful mentoring: a balance between intellectual freedom and guidance, consistent availability and involvement, resources for fostering development, positive feedback that is specific, treatment of graduate students as respected collaborators, and individualized attention to the student.[11]

They quote one mentor who said that mentoring "involves trying to help develop values—the values necessary to become a successful scientist—curiosity. So you mentor [students] by listening to them. You mentor them by encouraging them. You mentor them by showing them how you respect their individuality and autonomy." A Harvard zoologist and evolutionary geneticist described how his own mentor provided formal and informal opportunities for him to meet leaders in the field. "Hobnobbing with bigwigs in the field, he became a 'famous graduate student.' "[12]

Richard Tapia has said: "The post-doc position may be the most critical step in either making or breaking a successful future in the academy." He adds: "Graduate research advisors must take a role in finding a strong post-doc position for students with potential . . . graduate advisors must elicit a commitment of that kind of relationship from the post-doc advisor and then check to see that it is happening."[13]

Joseph Epstein collected essays from leading scholars and intellectuals about the professors who had influenced them the most. The essays originally appeared in the *American Scholar*, a journal edited by Epstein. In his introduction to the book, *Masters: Portraits of Great Teachers*, he says:

> Great teachers have left no record of their pedagogical accomplishments. The effect of their work has been rather like that of opera singers before the advent of recordings: there was, that is to say, no trace of their work beyond the circle of their audience. It does not do to overemphasize the comparison, but there is a sense in which teaching, like opera, is a performing art. Not only must the teacher keep up with his subject, but he must get it across. . . .
>
> What all the great teachers appear to have in common is love of their subject, an obvious satisfaction in arousing this love in their students, and an ability to convince them that what they are being taught is deadly serious.[14]

Among these great teachers was Alfred North Whitehead, whose best-known pupil was Bertrand Russell, who said: "Whitehead was extraordinarily perfect as a teacher. He took a personal interest in those with whom he had to deal and knew both their strong and their weak points. He would elicit from a pupil the best of which a pupil was capable. He was never repressive, or sarcastic, or superior or any of the things that inferior teachers like to be."[15] Whitehead's wife, Evelyn, said: "When we first came to Harvard, Altie's [Whitehead's] colleagues in the department said, '*Don't let the students interfere with your work*!' Ten or fifteen minutes is long enough for any conference with them." Whitehead rejected this advice and committed large blocks of time to meeting with the students.[16]

Anthropologist Ruth Benedict wrote to her mentor, Franz Boas, toward the end of his life: "There has never been a time since I've known you that I have not thanked God all the time that you existed and that I knew you." Epstein notes that Benedict's work was quite different from that of her mentor: "It may be agreeable for a student to have a teacher whose views are the same as his or her own, but in that case nothing fundamentally new can be learned."[17]

Page Smith has decried the declining status of teaching in the university: "The point I wish to make is a simple one. There is no decent, adequate, respectable education, in the proper sense of that much-abused word, without personal involvement by a teacher with the needs and concerns, academic and personal, of his/her students. All the rest is 'instruction' or 'information transferral,' 'communication technique,' or some other impersonal and antiseptic phrase, but *it is not teaching* and the student is not truly learning."[18]

F. H. T. Rhodes describes Alexander Agassiz as a great mentor:

> Alexander Agassiz, the great nineteenth-century Harvard geologist and zoologist, . . . used to give each of his students a fish, with the instruction to observe it carefully. Then he retired to his office nearby, available if needed. After a week of scrutiny, the students would be invited to describe their specimens. Usually, Agassiz would smile and say in a kindly way, "That's not quite right." And back the student would go—to make more observations, sketches, measurements, hours of study, until at last, weeks or months later, the professor was satisfied: the student had become a competent observer.
>
> Agassiz was teaching about more than fish, about more, even, than techniques of observation. He was teaching about learning, about precision, about self-confidence, about the progressive synthesis of knowledge and the satisfaction of discovery. He knew that the best education involves a "drawing out," not a "forcing in," that the professor has to persuade students of their own ability, to instill not just competence but also self-confidence. How much of this benefit would have been lost if, like some of today's professors, Agassiz had given an "easy A" on the student's first effort?[19]

In 1996, the White House initiated the Presidential Awards for Excellence in Science, Mathematics, and Engineering Mentoring. These awards focus on mentoring of historically underrepresented groups (HUGs) in higher education. Each year since 1996, the award has been given to at least 10 individuals and 10 institutions. The recipients are honored in a special White House ceremony, and each mentor receives a $10,000 grant.

A comprehensive study of the recipients of the extraordinary mentoring award was conducted by Priscilla Gayle Harris Watkins in 2005. She carried out a questionnaire survey of the entire population of individual recipients and interviewed a number of them. Dr. Watkins found that "exemplar mentors" are those who "focus on retaining HUGs in (but not recruiting them to) STEM disciplines; practice a precise definable method of mentoring; follow an unwritten curriculum that teaches non-quantifiable variables about the discipline; [and] believe substance and quality of contact are more important than frequency. Furthermore, they consider the traditional 'one on one' mentoring model as obsolete."[20]

She also found that exemplar mentors were "internally driven" to mentor students, even when institutional support for mentoring was insufficient: "A recurring theme in the data was around the notion of mentoring as a state of being. These mentors had internalized the practice of mentoring to the point that they did not recognize it as a task or an expectation of the workplace. Mentoring for this group was a fiber of their being, as essential to them as is their passion for the discipline" (110).

Dr. Watkins discussed "a detailed map for success" in the mentoring relationship. Among the strategies practiced by the "exemplar mentors":

- They assume the protégé has the intelligence to master the subject.
- They do not assume that the protégé understands the nuts and bolts of how the academic system works, for example, how to add and drop a class: "Internalization of the discipline, how to write a resume, interview protocol, filtering information, and how to present at conferences are a few important hidden skills for potential scholars" (185).
- They empower students to make their own decisions; they do not make the decisions for the student. They do identify possible courses of action.

Dr. Watkins found that trust of the mentor by the protégé was critical in the development of the mentoring relationship. Such trust "enhanced the exemplar mentor's ability to build a bond with the protégé, thus allowing the relationship to move beyond the traditional teacher-student relationship" (169).

She concluded: "Exemplar mentors go beyond the call of duty to support and protect their protégés. They are readily available to their protégés during the most challenging points in their careers as well as their most successful endeavors" (185).

Ground-Breaking Research on Mentoring and Expectations

The transformation of mathematics education in America started with Uri Treisman. His innovations added new dimensions to the mentoring process and relied completely on high expectations. In the process, he showed us how to close the achievement gap. For these reasons, I describe his original research in detail.

As a PhD student, Treisman developed calculus workshop study groups in which minority students not only achieved, but excelled beyond anyone's expectations.[21] It is important to understand why the Treisman model works and what his results meant for the development of a talented technocracy in America. As we shall see, there are two crucial components to the Treisman model:

1. Students are expected to excel and to do extra, more difficult homework problems. This is *not* a remedial approach.
2. Students study in groups, with mentors, in addition to their individual study.

Perhaps this model is most noteworthy for the theories it rejects. The students who succeeded in Treisman's workshops and in many subsequent workshops across the country were underrepresented minority students, often from poor or disadvantaged backgrounds. Their success in the workshop environment sometimes surpassed that of Anglos and Asians studying the same subject at the same institution. Thus, Treisman's data provide yet another basis for rejecting the notion that some students, because of gender or ethnic background, are less capable of mastering calculus—and by implication, other mathematical and scientific disciplines.

These findings also refute the notion that growing up in poverty or in an environment with a high incidence of neighborhood violence, dysfunctional families, health problems, and the like rules out a technical career, or that attending a less respected elementary or secondary school with limited resources rules out success. Certainly, growing up in poverty, with the associated social and family problems, makes it much more difficult for a young person to succeed in technical fields. But it can be done, and the benefits to the individual, his or her family, and society are almost incalculable.

The Berkeley Calculus Workshop

Treisman's original research is vitally important and should be examined in depth. It is presented in its most detailed form in his 1985 doctoral dissertation, "A Study of the Mathematics Performance of Black Students at the University of California, Berkeley." Treisman observed at the outset that he "had come to question the efficacy of individualized tutoring, self-paced instruction, and short courses aimed at the development of study skills—the traditional pedagogical arsenal of special programs for minorities. . . . These programs were remedial underpinnings: they focused on minority students' weaknesses rather than on their strengths" (2). He continues:

> In the Fall of 1975, while developing a training program for Mathematics Department teaching assistants, I became aware of the high rate at which Black students were failing freshman calculus at Berkeley. I had made it my practice to speak with the T.A.s about the weak and strong students in their teaching sections, and the regularity with which Black students appeared within the former group and Chinese students within the latter struck me as an issue that should be addressed in the training sessions. With this in mind, I began to seek the reasons for this apparent difference in performance. (4)

Treisman then interviewed 20 African American and 20 Chinese students, asking them about their use of instructors' office hours, whom they studied with, how much they studied, and so on. In addition, he asked the students to prepare a report, from memory, of how they had spent their time during the three days preceding the interviews. So far, these are standard research techniques. But Treisman added a wrinkle that produced startling results: he asked permission to accompany the students while they studied. "Many agreed, and over a period of eighteen months I accompanied these students to the library, their dormitory rooms, and their homes in the hope that I might see first-hand how they went about learning and doing mathematics" (5).

Obviously, the students' behavior was not unaffected by the presence of a mathematics instructor looking over their shoulder while they studied. "The students typically responded to their discomfort by asking me questions," noted Treisman, "and I to mine by rolling up my sleeves and helping them with their homework" (5). While this interaction would conflict with the demands of a rigorous experimental design, it facilitated the communication that was so important for Treisman's work in this heuristic, or hypothesis-generating, stage. Under these circumstances, the students really opened up.

Treisman presented some case study narratives about the experiences of several students to illustrate the challenge. Here are some excerpts from his narrative about a student named Joe.

> Joe graduated from a predominantly Black high school in East Oakland.
>
> Few students from this high school went on to college, and most of those who had come to Berkeley had been academically unsuccessful.

> Joe was president of his Junior class and was interested in school affairs, but he felt that preparing for college required that he hold himself apart from his classmates.
>
> Joe was actively recruited by Berkeley (his high school GPA was above 3.8), and was admitted directly into the College of Engineering.
>
> Joe's first few weeks at Cal were difficult. He had trouble understanding his professors' lectures and was often unable to complete his homework. When he received a D on both his calculus and computer science midterms, he was stunned. He believed he was well prepared for the University, and he knew he was trying hard to excel.
>
> Upon learning at the end of the term that he had failed both Chemistry and Computer Science—the final examinations contained few questions on material that Joe had reviewed—he decided to refrain from all social activity and to devote even more time to study. This new regimen, however, had little effect on his performance. After his winter quarter midterms, Joe stopped going to class; by then he was depressed and believed no matter what he did, no matter how hard he tried, he would fail his courses. (6–7)

Joe withdrew from Berkeley at the end of the academic year.

Treisman also gave a case study example of an African American student from a predominantly white high school who experienced difficulty adjusting to the university. Treisman observed that "among Black students in the Berkeley 1975 freshman class, only two of the twenty-one students who had enrolled in first-term calculus (Math 1A) went on to complete the final course in sequence (Math 1C) with a higher grade than C" (11).

What did Treisman find when he compared the study habits of Chinese and African American students? First, African American students almost invariably studied alone. In contrast, most of the Chinese students studied with other Chinese students. This gave them support and allowed intellectual exchanges; they tried to work out homework problems and the like, and they shared grapevine information about the course and the university. For example, they discovered that the professors' expectation of two hours of study for one hour of class was a serious underestimate. The Chinese students averaged 14 hours of study per week for a four-unit class, while the African American students had been devoting the eight hours per week that the professors recommended. The Chinese students critiqued one another's work, correcting errors and suggesting innovative solutions.

Compounding the performance discrepancy was the tendency for African American students to avoid remedial tutoring programs. These were students who had excelled in high school, many of them valedictorians, and they identified such programs with low-achieving students. They rarely approached their teaching assistants, "but, even when they did, their inability to define their needs clearly coupled with the low expectations that many T.A.'s held for the academic achievement of Black students usually precluded a fruitful exchange" (19).

Treisman designed a workshop to confront the problems his observational research had identified. Following a short pilot, the workshop program began in the fall of 1978. Participation was voluntary, and the students typically met for two hours a day, three or four days each week. The results were eye-opening: "Only one of the forty-two participants in the 1978–79 workshop failed the calculus class, and more than half of the students received grades of B– or better" (29). Following this initial success, the program was expanded, and external funding was obtained in the form of a three-year grant from the Fund for the Improvement of Post-Secondary Education. The program grew rapidly—within four years over 300 students were participating—but Treisman noted that the guiding philosophical principles remained the same:

> 1. the focus on helping minority students to excel at the University, rather than merely to avoid failure;
> 2. the emphasis on collaborative learning and the use of small-group teaching methods; and
> 3. the faculty sponsorship, which has both nourished the program and enabled it to survive. (30–31)

The program would begin with an orientation session: "Throughout the orientation, students are warned of the dangers of studying in isolation from their peers. They are encouraged to seek out classmates who share their high standards and goals, and to form study groups—even if they do not wish to participate in [the program]" (38). Since students' success seemed to be so closely tied to workshop participation, Treisman's description of those sessions assumes particular importance.

> A visitor entering a workshop session might mistake it at first for a noisy study hall or a lively math club meeting. Many of the students are engaged in discussion, some huddled in groups of three or four, some in pairs. Several students appear to be joking; others are working on a problem by themselves, ignoring the buzz about them.
>
> But after observing for a while, the visitor perceives the organization of the workshop more clearly. The grouping of students is transient; the small clus-

> ters form and reform to compare notes on the various problems that appear on a worksheet that has been distributed at the beginning of the session. Occasionally, a student will address the entire group, asking in a loud voice if anyone has solved a particular problem.
>
> Meanwhile, the workshop leader circulates unobtrusively, observing the students at work on the problems he has chosen for them. From time to time he sits down a short distance behind one of the groups and listens in for a few minutes. He might move on without addressing the group at all, or he might join in the discussion. Quite often, after working with a group, the leader takes one of its members aside for a brief period of private instruction.
>
> On occasion, the leader stops the proceedings and addresses the group as a whole. He might comment on a group's work, discuss one of the more perplexing worksheet problems, or ask one of the students to present his work to all in attendance. Such interruptions typically account for about twenty minutes of the two-hour session. During the remaining time, it is the students who seem to be in control; each is free to choose the problems on which he works and the students with whom he discusses his results. (41–42)

The workshop leaders spent considerable time preparing for the sessions; most of that time was devoted to creating problems. These included problems like those on tests, problems designed to reveal individual student deficiencies, problems that illustrated major course concepts, problems to provide experience with mathematical symbolism and language, and "street mathematics," that is, computational shortcuts.

Treisman notes that the leader's role was not to duplicate the instruction provided by the teaching assistants: "The leader's role is more like a manager or moderator than an instructor. *The basic premise of this workshop instruction is that through the regular practice of testing their ideas on others, students will develop the skills of self criticism essential not only for the development of mathematical sophistication, but for all intellectual growth.* Moreover, by continually explaining their ideas to others, students acquire the same benefits of increased understanding that teachers themselves regularly experience" (45–46; emphasis added).

Students did not participate in the workshop during their sophomore year and were discouraged from studying there. However, they were not simply dumped, unprepared, into deep water after their freshman year, a year in which virtually all of them had been successful at calculus. Rather, considerable attention was given to providing them with the skills and resources that would carry over into their subsequent undergraduate studies. For example, they were encouraged to maintain the network of friends developed during the workshop sessions and to become actively involved in other campus organizations.

When the calculus grades achieved by African American workshop participants were contrasted with those of African American students who did not participate in the workshops for the years 1973–84, the principal finding was that "the average grade earned by Black workshop participants has been approximately one full grade higher than that earned by Black students not participating in the program" (62). More than half of the workshop students received a B– or better in Math 1A, compared with fewer than one-fourth of the non-workshop students (64). Additional analysis revealed that the workshop students consistently outperformed non-workshop students in the courses following Math 1A—for example, in the second- and third-term calculus courses.

African American workshop students were considerably less likely than African American non-workshop students to drop out of school in the two years following this calculus course, that is, at the end of two and a half years of college. Furthermore, workshop students were far more likely to graduate from college than were non-workshop students. The graduation rates for workshop students, for example, were roughly comparable to those of nonminority students at Berkeley, while graduation rates for non-workshop participants were substantially lower. Finally, African American workshop students were more likely to persist in a mathematics-based major until graduation than were African American non-workshop students. *When SAT level was controlled—that is, when students who entered Berkeley with the same SAT level were compared—the African American workshop students outperformed all other groups in Math 1A, including the white and Asian students!*

Building on the insights and achievements of this early work at Berkeley, Dr. Treisman has engaged in a number of programs and projects to improve mathematics and science education. He has served on high-level national policy boards focused on mathematics and science education, including a special commission created by the Carnegie Foundation and the Institute for Advanced Study.[22]

Colleges and universities where the model that Uri Treisman developed at Berkeley has been applied include

- the University of Texas–Austin, in the Emerging Scholars Programs;
- Rutgers, the State University of New Jersey, in Project EXCEL;
- City College of New York;

- California Polytechnic State University at San Luis Obispo; and
- the University of California at San Francisco School of Medicine.

Presently a professor of mathematics and public affairs, and executive director of the Charles A. Dena Center at the University of Texas, Dr. Treisman has been honored repeatedly by his professional peers for his groundbreaking research; for example, he was named 2006 Scientist of the Year by the Harvard Foundation of Harvard University.

Twenty-five years ago, Uri Treisman demonstrated how mentoring and high expectations could dramatically improve the academic performance of students who had seemed destined for failure. But would these strategies be as successful in other kinds of institutions—for example, in less selective colleges and universities? More generally, do we now possess the strategies, knowledge, and skills to eliminate the achievement gap between advantaged and disadvantaged students? The next chapter addresses these questions.

CHAPTER 6

Closing the Achievement Gap

> Calculus is the mathematical study of change. Its essence is best captured by its original name, "fluxions," coined by its inventor, Isaac Newton. The name calls to mind systems that are ever in motion, always unfolding.
>
> Like calculus itself, this book is an exploration of change. It's about the transformation that takes place in a student's heart, as he and his teacher reverse roles, as they age, as they are buffeted by life itself. Through all these changes, they are bound together by a love of calculus.
>
> Steven Strogatz, *The Calculus of Friendship*

In this chapter, I review how the powerful combination of committed mentors and high expectations has closed the achievement gap at a variety of institutions. We will examine data on the academic progress of disadvantaged students and study the strategies that contributed to their success.

Despite the impressive student achievements, a reader of the original Treisman Berkeley report might have asked whether these workshops would be effective at other institutions. After all, the University of California at Berkeley is one of the top universities in the country, and students who are accepted there, including minority students, have already demonstrated that they are outstanding. Also, would these same effects be observed among students from other underrepresented minority groups—for example, Latinos?

The Calculus Workshop Programs at California State Polytechnic Institute, Pomona

One of the most extensive applications of the calculus workshop model has been at the California State Polytechnic Institute in Pomona. Martin Bonsangue conducted a thorough, systematic evaluation of this program, in which the majority of participants were Latinos. The evaluation focused upon a sample of 133 workshop and 187 non-workshop Latino American, African American, and Native American students, who were followed for a five-year period.[1] Both the treatment and comparison groups consisted largely of Latino students (87 percent of the workshop students and 85 percent of the non-workshop students were Latino). Both groups also contained mostly men (74 percent of the workshop students and 80 percent of the non-workshop students).

The workshop program was patterned after the Berkeley experience. Students met in groups to work on calculus problems twice a week for two hours each session. Statistical comparisons revealed no significant differences between the workshop and non-workshop students in terms of their background and other pre-intervention variables such as SAT scores, high school grade point averages, and scores on a precalculus diagnostic test. Further analysis revealed that each of several workshop subgroups—Latinos, African Americans, and women—outperformed the corresponding non-workshop group.

In the course of this research, Bonsangue made an important discovery. After many long hours sorting through paper forms (full course data were not available electronically at that time), he discovered that many students were failing and then retaking calculus again and again—sometimes four or five times. He then developed the course attempt ratio (CAR), in which the numerator is the number of times a student attempted a course and the denominator is the number of courses the student completed successfully (5).

Additional analyses showed that the workshop effects persisted in subsequent second-year calculus courses. The CARs of groups in the second-year courses did not differ, but the workshop group successfully completed one and a half times the number of second-year mathematics courses that the non-workshop group did, even though there were initially 40 percent fewer students in the workshop group than in the non-workshop group (8).

Workshop students were significantly less likely to drop out of the institution. In the 1986–89 sample, fewer than 4 percent of the workshop minority students dropped out by spring 1991, compared with 42 percent of those not in the workshops (9). Even more impressive, "of those students still enrolled in Mathematics, Science, and Engineering, more than ninety percent of the workshop students had completed their mathematics requirements for their individual majors, compared to less than sixty percent of the non-workshop students" (10). And workshop students who persisted in their MSE field achieved higher grade point averages overall than did the non-workshop students. However, they held only a slight advantage in terms of grades within the major, and there were no differences between the two groups with respect to number of units completed overall or within their majors.

In a further assessment of the relative importance of precollege academic measures and workshop participation, two multiple regressions were run predicting, first, MSE persistence and, second, mathematics completion:

> None of the precollege measures entered the regression equation for either dependent variable. Thus, the traditional pre-college cognitive measures held minimal power in predicting persistence in an MSE major and mathematics completion. *Workshop participation was the only statistically significant predictor of mathematics completion*, accounting for twenty-three percent of the variation in mathematics completion among men, while nearly twice that amount, forty-four percent, among women. . . .
>
> An additional aspect of this analysis was to examine the calculus achievement of minority workshop and non-workshop minority and majority students, with workshop students achieving a grade mean nearly one full grade point higher than that achieved for non-workshop minority students. *In fact, workshop students earned a calculus grade mean of at least .75 grade points higher than any non-workshop ethnic group, including Asian and white students, even though some precollege cognitive factors for the workshop group were significantly lower than those for white and Asian groups.* (15–17; emphasis added)

Additional regression analyses showed that the only significant predictor of calculus grades for workshop students was the number of hours spent in individual and group study, which accounted for 26 percent of the variation in course grades. Interviews with the students from the 1987–88 cohorts revealed that about half formed study groups in their upper-division courses after completing the workshop program, a finding that parallels the goals and experiences of the Berkeley program. The following interview finding is relevant to the discussion in chapter 2 of women's experiences in college science and mathematics courses: "Most of the women interviewees reported feelings of isolation or self-doubt to varying degrees, despite earning grades as high or higher than those of their male colleagues. Several women openly discussed barriers of sexism that they had experienced within their majors. Overall, women described a college experience that was qualitatively different from that described by men" (19).

Bonsangue conducted an additional analysis to assess the financial implications of these research findings. He found that the per-student cost-effectiveness (c.e.) for an intervention program is equal to the reduction in CAR ($tocar) times the cost to the university when a student takes a course (univcourse) minus the cost to the university of conducting a workshop (wkcost). This relationship can be summarized as:

$$\text{c.e.} = (\$\text{tocar} \times \text{univcourse}) - \text{wkcost}.$$

Financial data indicated that the cost to the university for a four-unit, one-quarter calculus course was $517. Figures from the Academic Excellence Workshop Budget Report indicated that the total direct and indirect costs of that effort were $335 per student. Thus, the cost effectiveness per workshop student in 1990–91 was

$$\text{c.e.} = (1.01 \times \$517) - \$335,$$

or $187 per student. Thus, using the difference in course attempt ratios from the 1986–89 cohorts (the mean CAR for non-workshop students was 4.64 while the mean CAR for workshop students was 3.63), students who participated in the workshop each cost the institution $187 less than students who were not in the workshop (22). Moreover, the savings to the institution may be even greater when subsequent calculus courses are considered and when the persistence and retention rates are considered.

Implementing workshop programs saves money for institutions and for students.

In summarizing the implications of the data yielded by the Pomona calculus workshop program, the evaluator stated: "The data strongly suggest that achievement among under-represented minority students in mathematics, science and engineering disciplines may be less associated with pre-college ability than with in-college academic experiences and expectations" (23).

Note the similarity of this conclusion with Lisa Loop's findings about the CGU Teacher Education program discussed in chapter 4. In short: *Effective programs make a difference. Pre-college variables, such as growing up in poverty, are not statistically significant.*

The work of Treisman and Bonsangue demonstrated that minority students—who traditionally have been underrepresented in science, mathematics, and engineering—can excel in these disciplines. Many of these students enter college with enormous social handicaps, yet they can achieve in calculus and subsequent courses. In fact, with the aid of the workshop program, they achieved at levels exceeding those reached by Anglo and Asian students. The key philosophical components of these workshop programs are (1) an emphasis on academic excellence rather than remediation and (2) mentoring facilitated by the formation of study groups.

The Harvard Assessment Seminars

Surprisingly parallel research findings have also come from the other end of the academic spectrum, on how super-achieving students learn best at Harvard College. In the late 1980s Richard Light conducted extensive explorations with students and other faculty in the Harvard Assessment Seminars. Light's main finding was that "students who get the most out of college, who grow the most academically, and who are happiest, organize their time to include interpersonal activities with faculty members, or fellow students, built around substantive academic work."[2] Light added that this is difficult for some students.

Another principal finding from the Harvard Assessment Seminars related directly to science. Noting that "more than half of Harvard freshmen express a strong interest in doing some work in science and . . . nearly half plan to concentrate in the sciences," Light described how this group of freshmen divided into two subgroups during the undergraduate years. The members of one subgroup loved their science experience and planned to study and work more in science. The other subgroup found their science courses dull and lost interest in a technical career. What accounts for the difference? Light concluded that it resulted directly from how their professors organized their science courses: "When asked to describe how they approach their work, students from these two groups sound as if they are describing different worlds. Those who stay in science tell of small, student-organized study groups. They meet outside of formal classes. They describe enjoying intense and often personal interaction with a good lab instructor. In contrast, those who switch away from the sciences rarely join a study group. They rarely work together with others. They describe class sections and lab instructors as dry, and above all, impersonal."

Science professors who succeeded in structuring their classes and labs to help undergraduates work collegially were honored by students and mentioned repeatedly. The word "inspiring" was used often. These professors attracted specialists in both sciences and other disciplines to their courses. Their success was not due to some mysterious charisma. It was due to the way they organized the work in their courses (10).

These findings about Harvard College students parallel the results of a doctoral dissertation study I had completed there.[3] In that research, I hypothesized that student interaction with reference groups—faculty, other students, family—would have an important impact on several key outcomes in the undergraduate experience, including academic achievement. I analyzed extensive longitudinal data on a sample of undergraduates in the Harvard classes of 1963 and 1964 using multivariate statistical methods—that is, stepwise multiple regression and canonical correlation analyses. The data had been collected as part of an ongoing research project, the Harvard Student Study, and included a variety of psychological and sociological measures. Each student in the sample devoted several days to the project every semester in college. The results showed the powerful effect of reference groups, especially faculty members. Those students who interacted with faculty members got substantially more out of the undergraduate experience, in terms of such measures as academic achievement and satisfaction, than did their contemporaries who worked as hard but failed to initiate such contacts.

Specific parallels between the observations from Light's Harvard assessment and those from the Treisman and Bonsangue research are striking. Consider Light's comment about how undergraduate learning can be made more effective.

> The students' second suggestion for interpersonal engagement is not unanimous. Just over half bring it up. It is the value of forming small study groups, with fellow students, that meet outside of class once or twice a week. The typical size of such groups is four to six students. They meet for an hour and a half to two hours. They nearly always include both men and women. Students stress that these are not intended to be "men's groups" or "women's groups"—they really are designed to be *study* groups. Those who participate in such groups take them very seriously.[4]

While the Harvard Assessment Seminars yielded these very clear recommendations about student participation in study groups, they also yielded observations about student involvement in campus life generally. One of the early findings was that there is "a strong connection between academic success and becoming an active member of the college community. The evidence is overwhelming that students who are active in extra-curricula activities adjust far more quickly to college life. And their grades are at least as high as those for students who concentrate narrowly, and only, on course work" (43). This observation mirrors both the theories and the research findings of such authors as Alexander Astin, Ernest Pascarella, William Spady, and others. Perhaps the most consistent finding in the higher education literature about the impact of the college experience on undergraduates has been that academic and social involvement with the campus community leads to success.

Another of Light's findings parallels an observation from the research at the California State Polytechnic Institute at Pomona: "While the advantages of study groups are widespread, there is one group of students for whom they seem especially important: young women concentrating in the physical sciences." Light cites the 1988 undergraduate honors thesis of Andrea Shlipak, who found that "women who concentrate in physics and engineering consider these small working groups a crucial part of their learning activities. Further, her interviews with women who enter college intending to specialize in the physical sciences reveal a sharp break between those women who join study groups and those who don't. Women who join a small study group are far more likely to persist as science concentrators than those who always or nearly always study alone" (54).

In a section about the value of building substantive work in science around more student-student and student-faculty interactions, Light observes:

> This is hard to do. To some it will sound vague. Yet this suggestion is brought up more than any other by students. Many perceive serious work in the physical sciences as impersonal. In contrast, they think of classes in humanities and social sciences as "dealing with people—their dilemmas, their joys, their tragedies, their lives," as one woman who switched from chemistry to anthropology put it.
>
> When I shared this point with faculty colleagues in the sciences, one responded, "But physics and chemistry and biology are beautiful, too—just in a different way." I know. And some students know, too. But unless professors make a conscious effort to share their perceptions of this beauty, they will continue to lose some students. And the most promising way to share such perceptions, according to students who have chosen to work in the sciences, is to build small work teams so students interact more. For example, create a discussion group after each major lab experiment. That way, rather than going home alone into the night, each student can share findings, frustrations, and surprises with others. They become part of an ongoing conversation shared by young fellow scientists. (67–68)

Research in three different institutional environments—UC Berkeley, California Polytechnic Pomona, and Harvard—demonstrated the power of expecting students to excel and the effectiveness of mentoring via workshop study groups to help them achieve success.

Mentoring and high expectations are crucial to closing the achievement gap. Successful mentoring programs today often build directly upon Treisman's early research.

Closing the Achievement Gap in Louisiana and Texas

The Louis Stokes Alliance for Minority Participation (LSAMP) is a program that awards grants to consortia of colleges and universities, typically $5 million for the first five years, to broaden participation in undergraduate and graduate STEM education. The program is available to STEM students of all ethnicities.

The person with the original vision for this program was Luther Williams, of the National Science Foundation. LSAMP's current director is Art Hicks. I have worked for many years with LSAMP in Louisiana and in Houston. Both programs have had a strong impact on the STEM workforce in their states.

Fitzgerald Bramwell and Elinor Brown, who have analyzed data about STEM doctorates from the NSF, have traced the baccalaureate origins of PhDs over time and have found that the LSAMP program has a clear impact. For example, they found that the number of PhDs awarded to college graduates from Louisiana has increased dramatically.[5] For some time, federal STEM programs in Louisiana had been coordinated impressively at the state level. In 1995, the implementation of the LSAMP program focused these efforts on STEM bachelor's degrees. There were concomitant increases in the rate of bachelor's degree production in STEM disciplines beginning in 1999. The timing of these increases is consistent with the later completion of doctoral studies by the first students whose undergraduate work had been supported by LSAMP. These sharp increases in doctorates are most evident in women and African Americans who earned their bachelor's degrees in Louisiana.[6]

Each year for several days I visit Southern University as part of an LSAMP evaluation team with my colleagues Fitzgerald Bramwell and Michael Howell. The leader of the Louisiana LSAMP program, Professor Diola Bagayoko, a physicist, has developed a highly effective mentoring academy. It is instructive to look at his accomplishments in inspiring students.

A World-Class Mentor in Baton Rouge

In the 12th century, the University of Timbuktu flourished on the banks of the Niger River in the African nation of Mali. With roughly 25,000 students, the university drew faculty from around the world. One of the most respected faculty members was Ahmed Bagayoko.[7]

Roughly 800 years later, physicist Diola Bagayoko, who originally is from Mali, created a mentoring academy at Southern University in Baton Rouge, Louisiana. The historic African university was the intellectual inspiration for this academy. In fact, Dr. Bagayoko named the mentoring academy, which has become a national model in the 21st century, the Timbuktu Academy. Dr. Bagayoko reflects: "As a young man in Mali, I was constantly reminded of the accomplishments of Professor Bagayoko. Being a Bagayoko inspires great challenge!" (48). He adds, "Southern University Baton Rouge is located on the left bank of the mighty Mississippi the same way the city of Timbuktu is on the left bank of the majestic Niger river" (51).

When Professor Bagayoko, a master mentor, reflects about his own intellectual development, he credits a series of mentors including Robert Verdier.

> From the 7th through the 9th grade, I met Mr. Robert Verdier, a young French man who taught me French language. As a result of his exceptional teaching, coupled with the weekly essays that guaranteed that students actually learned, I basically completed my learning of the French language by the ninth grade. Mr. Verdier, as my mother told me, was not just my teacher; he was also my mentor. Indeed, while I was in the eighth grade, he gave me the entire collection of the work of Victor Hugo! Victor Hugo, for those who might not know, is the most celebrated writer in the French literature. Naturally, while I repeatedly said "thank you" to Mr. Verdier, I knew that the real "thanks" was only to come after I read every single one of those masterpieces. My vocabulary and my overall mastery of French literally took off. (48)

Dr. Bagayoko is a respected and productive physicist, with more than 70 publications in physics. He and colleagues developed the Bagayoko, Zhao, and Williams (BZW) method

> to resolve a close to one century old problem stemming from a woeful underestimation of band gaps of semiconductors by previous theoretical calculations. This method is expected to have profound implications in condensed matters. It is almost unavoidable in the theoretical studies of nanostructures that are currently attracting great interest. Further, the BZW method, when applied to the shell model of nuclear physics, is the only computational method known to us that can provide accurate gaps between the occupied and the unoccupied nuclear energy levels. These accurate gaps between the occupied and the unoccupied nuclear states are expected to lead to promising explorations of gamma ray amplification by stimulated emission of radiation (i.e., graser). (51–52)

Dr. Bagayoko also has about 40 publications focused on mentoring and learning, publications that provide the intellectual foundation for the successful academy he has created at Southern University Baton Rouge. In the Timbuktu Academy, Dr. Bagayoko and his associates apply the "power law of performance":

> The power law of performance of practice (PLP) is perhaps the most stable of the laws in cognitive science. It states that the time (T) it takes an individual to perform a given task decreases as the number of times (N) the individual practices the task increases. In mathematical terminology, the power law is
>
> $$T = A + B(N + E)^{-p} \text{ or } T = A + B/(N + E)^{p}$$
>
> where A, B, E, and p are constants that vary with the task at hand and with the individual performing the task. The constant A represents a physiological limit. The constants B and E partly denote prior experiences before the beginning of the practice sessions, and p is the learning rate. *In essence, PLP states that practice renders perfect.* This law applies to the performance of sensory-motor (or athletic), creative (or artistic), and cognitive (or intellectual) tasks. The shorter the time taken to perform the tasks—completely and correctly—the higher the level of proficiency of the individual. Hence, as the number of practices increases, so does the proficiency of the individual.[8]

The systematic mentoring model of the Timbuktu Academy emphasizes the following 10 strategies or activities:[9]

1. Financial support is necessary for optimum student performance.
2. The enhancement of communication skills is paramount.
3. Comprehensive, scientific advisement is a necessity.
4. Tutoring is not just for remediation.
5. Generic research activities include rigorous literature searches.
6. Students are required to undertake specific research projects.
7. Scholars are immersed in a professional culture.
8. The development and enhancement of computer and technological skills are essential.
9. Monitoring throughout the semester prevents potential problems.
10. Guidance to graduate school begins in the freshman year (or earlier).

The Timbuktu Academy mentors physics majors and graduate students as well as other students at Southern University and also mentors precollege students to facilitate and support their transition to college, especially to STEM majors. While the academy is open to students of all ethnicities, the majority of scholars in this academy, which is located in a historically black university, are African American. In the 15 years from 1994 to 2009, 161 African American students were mentored by the academy in physics (83), chemistry (29), and engineering (49). Of those, 95 enrolled in or completed graduate school (59 in physics, 19 in chemistry, and 17 in engineering).

In the 16 years beginning with 1994, the precollege programs of the Timbuktu Academy have enrolled 2,093 participants. According to a report provided by the Timbuktu Academy,

- Due in part to the college portfolio the high school students (SSI, Challenge 2000, and ESTA) prepared during the summer, practically 100% of these student participants attend college upon their graduation from high school.
- Approximately 80% of these summer scholars enroll in college science, technology, engineering, and mathematics (STEM) curricula.
- For the last three summers, the average gain in ACT scores registered by the Academy's high school scholars has been five to six times the average gain made possible by a typical high school during the same six-week period, according to ACT data.[10]

The Timbuktu Academy has received funding both from the NSF and the Defense Department Office of Naval Research. The journal *Science* reported in 2006: "On a campus that loses half of its freshman class, 90% of first year students who choose to major in physics—nearly all of them academy scholars—earn a degree in 4 years."[11]

Dr. Bagayoko, who exudes energy, enthusiasm, and humor, has received numerous prestigious awards for his mentoring accomplishments. He was one of the first 10 recipients of the award for excellence in science, mathematics, and engineering mentoring from the White House. He also was the recipient of a Mentor Award from the American Association for the Advancement of Science.

A few of his former students:

Anthony Pullen is to receive his PhD in physics in 2010 from the California Institute of Technology. His special area of interest is cosmology, including dark matter detection and the clustering of galaxies. His research can provide scientists with information about the average density of matter and of radiation in the universe. He aspires to become a professor in a liberal arts college, where he can contribute by teaching.

Pullen is originally from the Baton Rouge area. Dr. Bagayoko knew his family and reached out to this intelligent child in middle school. He secured funding for Pullen's undergraduate studies at Southern University and helped him find summer internships at the California Institute of Technology and the Massachusetts Institute of Technology. Pullen had two publications before graduating from college. He was accepted by MIT, Princeton, Stanford, and the University of California–Berkeley in addition to Cal Tech. Pullen says: "Dr. Bagayoko is very meticulous. He doesn't play games. He really is for the students. He sticks with his methods, sticks to his guns. He communicates his expectations to students very clearly. Those that follow his guidelines will excel because of his mentoring."[12]

Divine Kumah completed his PhD in applied physics at the University of Michigan in 2009. His doctoral dissertation was about the use of X rays to image the atomic structure of nanoscale systems. Presently he is engaged in postdoctoral work at Yale University. His goal is a teaching and research position in academia.

Kumah grew up in Ghana and completed his elementary and secondary education there, except for one semester as a foreign exchange student in El Paso, Texas. He entered college planning to major in engineering, but changed his mind, and his major, when he met Diola Bagayoko. He graduated from Southern University in three and a half years. Dr. Bagayoko continually stressed to him the importance of conducting research as an undergraduate, and while in college he secured research internships for three summers. He worked at the National Renewable Energy Laboratory in Colorado and then, for two summers, at the prestigious CERN laboratory in Switzerland. While there he built one of the detectors for the Large Hadron Collider. Dr. Kumah says, "Dr. Bagayoko makes you feel like your decisions are the only things that count in what you want to achieve. There are no closed doors."[13]

Other graduates include Casey Stevens, a PhD student at the University of Chicago and a recipient of an NSF Graduate Research Fellowship, and Dr. Zelda Gills, who earned her PhD at Georgia Tech and has published two papers on chaos in lasers in *Physical Review Letters,* the top journal in physics. Dr. Bagayoko mentored each of these students and also taught mathematical physics to each of them.

Closing the Achievement Gap in Houston

Martin Bonsangue and I have been external evaluators for the LSAMP program in Houston for a number of years. The Houston alliance includes eight institutions: the Houston Community College System, Rice University, San Jacinto Community College, Texas Southern University, Texas State University–San Marcos, University of Houston–Central, University of Houston–Downtown, and University of Houston–Victoria. In 1999 this consortium set the goal of doubling

TABLE 6.1
Minority students enrolled in STEM disciplines in Houston LSAMP programs, 1998–2004

Race/ethnicity	1998–1999	1999–2000	2000–2001	2001–2002	2002–2003	2003–2004
Black or African American	2,075	3,124	3,008	3,945	3,043	3,837
Hispanic or Latino	2,474	3,256	3,294	3,752	3,446	3,984
Native American	75	80	69	62	56	67
Native Hawaiian or Pacific Islander	34	74	38	0	91	305
More than one race reported, minority	7	5	5	0	7	27
All minorities	4,665	6,539	6,414	7,759	6,643	8,220

Source: Martin V. Bonsangue and David E. Drew, "Houston Alliance for Minority Participation: Ten-Year Evaluation Report" (report to the National Science Foundation, Feb. 2010).

the number of students of color who achieve bachelor's degrees in STEM disciplines in five years, and they came close to doing it. It's an astounding record. The question is, how did they do it and what can we learn from that?[14]

Table 6.1 presents data on minority enrollments in STEM disciplines in these consortium institutions for a baseline year, 1998–99, and over the first five years.

The NSF expects each LSAMP consortium to substantially increase the number of underrepresented minority students obtaining bachelor's degrees in the first five years, and ideally to double the number; it then expects each alliance to sustain this higher level of productivity during the next five years. Figure 6.1 presents degree data for the first 10 years of the Houston alliance.

The graph of minority STEM bachelor's degrees in the Houston consortium shown in figure 6.1 reflects two patterns, one overlaid on the other. First, there is a steady increase in the number of degrees awarded over time. Second, there is a "picket fence" effect, in which alternate years are either higher or lower. I have seen this picket fence effect in other STEM degree data—for example, in the graph of minority STEM degrees at McNeese State University in Louisiana. This pattern of alternating higher and lower productivity seems real. I suspect it may have to do with the availability of required advanced courses for STEM majors.

As you can see, these universities in Houston have demonstrated that you can close the achievement gap in a very short period of time. Dr. Bonsangue and I were interviewed by a reporter from a magazine published by the American Association for the Advancement of Science about the Houston experience

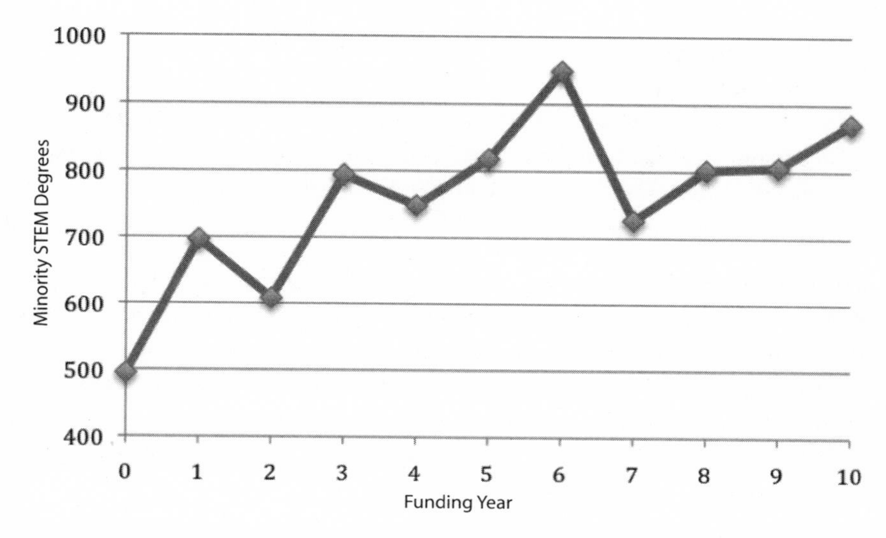

Figure 6.1. Undergraduate minority STEM degrees awarded in the Houston LSAMP program, 1998–1999 (Year 0) through 2008–2009 (Year 10). *Source:* Martin V. Bonsangue and David E. Drew, "Houston Alliance for Minority Participation: Ten-Year Evaluation Report" (report to the National Science Foundation, Feb. 2010).

not long ago, and the reporter asked, "How does this compare with the national growth rate?" When we retrieved data from the NSF Web site and made a longitudinal comparison, we found that the growth in degrees awarded to the Houston minority students in science and engineering was substantially more than the national growth rate in minority degrees.

- Assuming a linear increase at the rate of 20% per year, the expected number of minority STEM degrees awarded was 3,664. The actual number awarded was 3,499, or 95.5% of the expected number.
- Assuming a linear increase at the rate of 20% per year, the expected number of minority students enrolled in STEM majors was 27,990. The actual enrollment number was 27,355, or 97.7% of the expected number.

In fact, comparison against an expected growth rate of 20% is conservative. Because of compounding, doubling in five years would require only a 15% annual growth rate.

How have the universities in Houston accomplished this? They have used four strategies. And any other college or university can apply these strategies:

1. extensive recruitment,
2. constant mentoring,
3. creating a peer culture of student support aimed at academic excellence, and
4. engaging the community colleges and tapping the tremendous talent of people, often from poverty, who begin their college education at a community college.

Try for a moment to look at these strategies from the student's point of view. It's difficult for many of us who have completed college to remember what it was like to be beginning college, let alone for those of us who are white to appreciate the barriers facing a student of color, or for a middle-class person to appreciate the barriers facing a student from poverty. To those students, to use the words of an old hymn, it must seem that the college education that lies ahead is a combination of dangers, toils, and snares.

RECRUITMENT. One of the institutions in the Houston alliance, the University of Houston–Downtown (UHD), is located in a poor neighborhood, with many high school students who assume that a college education is out of their reach. Dr. Richard Aló leads an effort to reach out to those students and make them realize that both college and a STEM career can be possible for them. He and his UHD staff have connected with the students as early as the seventh grade to present these possibilities. They have been creative in their communication and outreach, even employing a social worker as part of this effort.

MENTORING. At Texas Southern University, Dr. Bobby Wilson is the driving force behind the excellence of the instructional program in the sciences. He expects the best from his students. He and the LSAMP staff and faculty are all committed to extensive mentoring of students.

Dr. Wilson is a distinguished chemist who previously was an NSF program officer. He holds the Shell Oil Endowed Chair of Environmental Toxicology and is the L. Lloyd Woods Distinguished Professor of Chemistry. He served many years as university provost, and for a lengthy period as acting president of Texas Southern, yet he still found time to give undergraduates focused individual attention. His commitment to teaching and mentoring started at a young age. While a doctoral student at Michigan State University, he received a chemistry department Excellence in Teaching citation in 1975.

He devotes several pages in his curriculum vitae to discussing his research, teaching, and administration. The first sentence is not about his publications nor his service as university president, but is about students: "Much of my attention was devoted to analyzing the trends and factors that impact the performance of students on the state-mandated test, Texas Academic Skills Program (TASP)." Dr. Wilson is a visible presence in his lab. He constantly banters with students, communicating high expectations, joking with them, and motivating them. They can see his commitment to research and to excellence in academic science on a daily basis.

Another influential presence on the Texas Southern campus is Michelle Tolbert, the university's LSAMP program director. I recall riding on campus in a golf cart with her; every time she saw an LSAMP student she stopped, she knew them, and she reminded them, for example, that they had an assignment due that day in chemistry. Dr. Wilson recruited Tolbert from the business sector, and she has brought executive efficiency to coordinating and directing the LSAMP program. She is devoted to the success of every student and she brings boundless energy to this task. There are over a hundred LSAMP scholars at Texas Southern, but they receive constant guidance and mentoring. They turn to Dr. Wilson, Ms. Tolbert, and committed faculty members, including Dr. Willie Taylor.

One Web site has referred to Professor Taylor as one of the "mathematicians of the African diaspora." Dr. Taylor was the first African American to receive a PhD in mathematics from the University of Houston. The title of his doctoral dissertation was "Oscillatory Properties of Non-Selfadjoint Fourth Order Differential Equations." Dr. Taylor's commitment to his students is deep and real—and it does not oscillate. He can be found tutoring students who are struggling with mathematics just about every day, all day—including weekends. In addition to the immense contribution he makes to Texas Southern students, we have observed that students from other Houston universities quietly come over to Texas Southern for his help in learning and understanding mathematical concepts.

At Texas State University, Dr. Salina Vasquez-Mireles was a key person in the early success of the LSAMP mentoring activities. In 2002 and 2005 she won the Presidential Award for Excellence in Teaching in the College of Science, and she has been nominated twice for the Mariel M. Muir Excellence in Mentoring Award. She is also an extraordinary role model for young Latinos and for students from disadvantaged backgrounds. She takes this role seriously.

In her scholarship, Dr. Vasquez-Mireles has identified and is addressing a critical issue: the crucial importance of language in facilitating or hindering the integration or correlation (the term she favors, for good reasons) of mathematics education and science education. When reviewing the dismal performance of American high school students on international assessments of academic achievement, some scholars have emphasized the need to better integrate presentation of math and science subjects in middle and high school in the United States. Vasquez-Mireles is the first scholar, to my knowledge, to focus directly on the ambiguities of words that have different meanings in math and science. I don't believe this is a subtle, fringe issue; this is profoundly important. She notes that the same words have different meanings in both math and science, while different words have the same meanings in both math and science. Examples of the first category include "regular," "average," "quarter," "element," "density," and "variance."[15]

CREATING A SUPPORTIVE PEER CULTURE. The University of Houston–Central sets aside four or five rooms where the students in the LSAMP program gather. It's a social experience, but it's all geared toward academic excellence. They don't just sit there and talk about the World Series. They are relaxed, but they're all working toward doing extra problems, towards excelling.

COMMUNITY COLLEGES. There is a tremendous pool of untapped talent at community colleges. Many people can't afford to go to college when they're 18 years old; then, when they later go to college, they still have staggering financial responsibilities and few financial resources. They often begin part-time at a community college. But, many of those students could become outstanding scientists and engineers. The Houston alliance has been working on articulation programs to facilitate the transition of these students from the two-year environment to a university.

Many of these students are Latino. Daniel G. Solorzano has published extensively about STEM education and Latino students. In one study he and Armida Ornelas examined conditions under which Latino community college students successfully transfer to four-year institutions. They report two dismal, related statistics: while 32 of every 100 Latino high school graduates enroll as freshmen in California community colleges, only 3 of those subsequently transfer to a university. In an intensive case study of a single institution, they then indentified barriers to transfer and presented recommendations for removing those barriers.[16]

A report by the Tomas Rivera Policy Institute notes that many post-secondary schools offer inadequate advising and mentoring programs for Latinos. The report states: "The colleges that are more successful are more likely to tailor support services in ways that leverage the strengths of Latino culture and family dynamics. For example, there are increasing examples of institutions successfully using peer and group-based support systems with Latino students."[17]

The pioneering work of Uri Treisman and the current success stories in Louisiana, Houston, and across the nation demonstrate that we can close the achievement gap and that mentoring and high expectations are essential in that effort.

This chapter has addressed the achievement gap between advantaged and disadvantaged students, which occurs in each level of education: elementary school, middle school, high school, and college. Another challenge facing STEM educators is the attrition of students from one level to another, especially in the transition from high school to college. The next chapter focuses on access to college and how that affects the STEM pipeline.

CHAPTER 7

College Access and the STEM Pipeline

The Case of Nevada

Nevada is ten thousand tales of ugliness and beauty, viciousness and virtue.

Richard Lillard, *Desert Challenge*

As students make their way through school, there are points when they may become less likely to study math and science—points where they may be lost forever from the STEM pipeline. One critical juncture is the transition from high school to college. We must make mathematics and science available and attractive to virtually every student in middle and high schools. If skilled teachers help students master these subjects, a college degree and a STEM career are realistic future prospects.

Unfortunately, many American students do not have access to affordable college education. Most colleges and universities are beyond the financial reach of many students and their families, leading them to borrow staggering amounts to finance their undergraduate educations. They then become modern indentured servants, struggling for years to pay off their college loans. Is a college education worth the expenditure of time, energy, and money?

Many young people choose occupations that do not require a college education. Some of the fastest growing jobs in our economy require skilled workers who need no more training than that available from a proprietary school or community college vocational program. Examples include wind tunnel technicians and underwater welders. Many people with an associate's degree earn as much as their counterparts with a bachelor's degree. And some graduates with four-year degrees have trouble finding jobs in the post-2008 recession economy.

However, when individuals entering the workforce can expect multiple jobs and careers over their lifetime, learning how to learn and to think critically—both staples of the undergraduate experience—have become more important than learning a particular occupation. College provides access to careers that are often not open to those without a college degree. The American dream has always given the poorest children the opportunity to succeed, and a college education is the main vehicle for moving from poverty to the middle class. The economic odds are against the high school graduate who does not continue his or her education.

A liberal arts college education leads to a better quality of life, even if lifetime earnings are not increased. A college education enlarges students' worldview, introduces them to culturally enriching experiences, and forces them to review, question, and solidify their core values. It essentially prepares a young person for survival and success in middle-class America, for leadership roles.

Access to college is a key issue in maintaining the STEM pipeline. President Obama wisely has recognized the important role that community colleges play as a port of entry into a four-year college education for many students. In 2009 Obama announced a multi-billion-dollar initiative to increase student access to community colleges.

Nevada is an ideal place to explore issues of college access. In a town outside Las Vegas, a new college was established to help increase access to higher education for thousands of previously underserved students.

What Happens in Vegas . . .

If one were to contrive a laboratory experiment to test whether a college education is worth the investment, it might look like Las Vegas. Consider the following:

- The Las Vegas economy is mostly composed of service jobs in casinos and hotels, making service employment an attractive career prospect for high school graduates. The most coveted service position in hotels (perhaps aside from managerial positions) is valet parking.
- Valet parking attendants in Las Vegas derive their income largely from tips. Income from tips often is not reported to the IRS and therefore is not taxed. Las Vegas parking attendants make about $60,000 per year.
- First-year teachers in the Clark County school district, of which the city of Las Vegas is a part, earn roughly the same as entry-level hotel housekeepers.

Several years ago, I was asked to be the lead academic planner for a new four-year college in Henderson, a suburb of Las Vegas. More colleges are desperately needed there. Nevada ranks last—behind even Mississippi and Arkansas—in the percentage of high school graduates who go to college. The college had bitter opposition from opponents who argued that a high school graduate can work in a casino or a hotel and make a huge salary. Las Vegas is a Petri dish of the value of a college education, and it can show us the impact of a college on the economy of a city or region.

A Tough Place to Launch New Ventures

The creation of a new college in the Las Vegas valley was meant to transform both the educational landscape and the economic power of southern Nevada. Two earlier attempts at massive innovation in Las Vegas aimed to accomplish similar goals, but only one succeeded.

In the late 19th century, Joseph Smith and Brigham Young transported their new Mormon religion from upstate New York to the American West. They reached Utah and sent emissaries to Nevada to continue the westward movement. The Mormon group got as far as Las Vegas. In the face of the hot, dry, dusty, inhospitable desert environment, they retreated to Utah. It was Utah that became the base for the Mormons and today is the center and home of that rapidly growing religion. A few Mormons remained in Nevada, and the influence of the Mormon community is pervasive and powerful in Las Vegas. But the original idea to headquarter the Mormon religion in Nevada was a bust.

Contemporary Las Vegas derives its inspiration from a brilliant, creative, innovative entrepreneur named Benjamin Siegel, also known as Bugsy. He was a ruthless gangster with a fierce temper. He hated his nickname; no one who knew him dared to mention it in his presence. In those days, the postwar 1940s, what little gambling existed in the United States was controlled by a small group of mobsters from New York, including Meyer Lansky, Frank Costello, Lucky Luciano, and Bugsy Siegel. Their organization was known as Murder Incorporated. Ben Siegel was head of West Coast operations. In 1945 he scanned the dusty, arid landscape of Las Vegas and articulated a vision of a gambling and entertainment mecca with grand hotels on a scale and in a style previously unimaginable. With the financial backing of Lansky and Luciano, he opened the first Las Vegas megahotel, called the Flamingo.

All Las Vegas hotels since the 1940s have been influenced to one degree or another by the design decisions Bugsy Siegel made in regard to the Flamingo. Despite its grandeur, the opening of the famed hotel was not without complication. Its construction was delayed, there were cost overruns, and Siegel's associates became suspicious. And its casino lost money at first. In March of 1947, just months after opening, the Flamingo closed. It reopened in May. That spring, Meyer Lansky met with the other mobsters who had invested money in the Flamingo, costs that increased substantially with the overruns in the construction of the hotel. They came to the conclusion that Bugsy had to go. Bugsy Siegel was killed at his Los Angeles home on June 20, 1947.

Benjamin Siegel was reckless and dangerous. He was a mobster who lived on the wrong side of the law his entire life. He lived violently and died a violent death. But he was also a creative man with a vision. The form, structure, design, and success of today's Las Vegas all can be traced back to Bugsy. Ben Siegel thought big. Perhaps he was influenced by Bonetti's law of gambling: The less you bet, the more you lose when you win.[1]

The Mormon excursion and the construction of the Flamingo teach us three lessons:

1. It's tough to be an innovator in Las Vegas.
2. Seed money is crucial to success.
3. The alternatives to success are not pretty.

Michael Ventura observed about the Mormon and Mafia excursions: "Odd, isn't it, that such similar social structures were drawn to the same place in the same immense desert, and that each built its first adobe on what is, in effect, the same street."[2] As it turned out, the academic task force for Nevada State College held its meetings in a library on the same street.

A Plan for Nevada State College

A major suburb of Las Vegas, Henderson was in 1999 growing at a pace unmatched almost anywhere in the United States. The town leaders had concluded that Henderson would benefit enormously from establishing a college. Upon hearing the arguments advanced by civic leaders from Henderson, the state legislature awarded $500,000 in planning money to the Nevada Board of Regents. The regents then awarded a contract for the college to the City of Henderson.

The new college was a high-risk venture, although no one realized just how substantial the risks were at that time. In the case of Nevada State College, there were both powerful supporters and deeply entrenched, powerful opponents and enemies.

Beyond local politics, there are serious questions that must be addressed by those planning a new college. Given the proliferation of on-line degree programs made possible by the technology explosion of the past 10 years, opponents to new college construction question the need for more traditionally structured colleges. If young people can land high-paying jobs in the computer industry without a college degree, is there still a reason to seek an undergraduate education? If the answer to that question is yes, should these young people be asked to study the traditional canon of western civilization, to read the work of Socrates, Hobbs, Locke, and the like? How should an undergraduate education—especially the first two years of college, when students are fulfilling general education requirements—be designed for today's student?

The Las Vegas Context

The saga of Nevada State College cannot be separated from its unique state environment. Consider first the geography. Nevada consists of vast expanses of desert and mountains interrupted by a few highly populated areas, including Las Vegas and Reno. Its economy is unique, too; Nevada is a one-industry state. The economy and culture of Las Vegas revolve around gaming and the Las Vegas Strip, where tourists routinely drop millions of dollars each week in games like roulette, blackjack, and craps, where the odds always favor the house. No guest in a Las Vegas hotel can walk from one area to another without walking through a room of slot machines. As a statistician, I'm amazed that people are willing to invest and lose so much money in a system they know stacks the odds against them. But they do.

The gambling industry contributes both complacency and a sense of urgency to planning about higher education in Nevada. There is complacency because young people can easily find high-paying jobs in the hotels and casinos. But there is also an urgent need for a highly skilled work force if and when the global high-tech economy leaves Nevada behind, or gambling ceases to be so profitable.

The story of the founding of Nevada State College includes high aspirations and low drama. Characters included the U.S. secretary of education, the U.S. deputy secretary of education, the governor, two former governors and their spouses, the lieutenant governor, the attorney general, the chancellor of higher education, the board of regents, the mayor of Henderson, the mayor of Las Vegas, the speaker of the Nevada Assembly, the majority leader of the Senate, several publishers, the former commissioner of education in Texas, university presidents, and the head of the J. Paul Getty Trust.

Also in the cast of characters were a lounge singer, a chief of detectives, a former defense attorney for the mob, and a graduate student studying medieval history. As it happens, all of the people in this second list also are represented in the first list above.

Fifty years ago, the federal government tested nuclear weapons in Nevada. You actually could see the mushroom cloud from the hills surrounding Las Vegas, and people held "atomic bomb picnics." The creation of the new state college at Henderson was a social experiment in which dangerous political, economic, and educational chemicals were combined in a volatile mix. There was indeed an explosion, and the light from that explosion illuminated both American higher education and Las Vegas.

Where else but Las Vegas would they make the Atom Bomb a picnic? An honest-to-God picnic. From 1951, when the bright mushroom first bloomed in the desert north of town, to 1962, when some killjoy treaty drove the testing underground, the casinos sponsored picnic lunches to view the A-blasts.[3]

Las Vegas is of course located in the middle of the desert. Increasingly during my work with Nevada State College I thought a lot about mirages, because things in Nevada were never quite what they appeared to be. I sometimes wondered if the new institution should be named "Mirage State College."

Early Academic Planning

The struggle to launch Nevada State College and the unique social and political context of Las Vegas, the state of Nevada, and the growing suburb of Henderson contain many lessons for those planning colleges elsewhere. Its founding involved numerous players with different, sometimes conflicting, interests and many philosophical debates. The board of regents was at the center of educational and political debates about whether Nevada State College would live or die and about what its nature would be. Some regents were vocal, outspoken supporters, and others were determined to prevent it from opening. In the end, the board voted 8 to 3 in favor of proceeding. In addition to the board of regents, four other organizations were involved in the decision to start a new college: the office of the governor, government and civic leaders from Henderson, the state legislature, and the office of the chancellor of higher education.

In theory (say, the way a civics textbook would describe it), the process for starting a new college would have begun with a recommendation from the chancellor's office to the board of regents. The board then would have received funding for planning from the state legislature and would, in the planning process, have selected a site for the new college. In practice, while the chancellor's office did make a recommendation, and the board of regents did request funding from the legislature, the impetus for the college came from civic leaders in Henderson. The two key players in Henderson were Mayor Jim Gibson and state assembly member Richard Perkins, who subsequently was to become the speaker of the assembly. Mr. Perkins also served as the chief of detectives for the city of Henderson.

In February 2000, the university and community college system of Nevada held a series of open meetings to gather opinions from the higher education community about the proposed new college. There was a wide range of responses. A proponent said: "I believe that the taxpayers of Southern Nevada desire a solid four-year college for their children. If you take a survey of the residents, I am sure that they will agree. I have lived in Henderson since 1946 and offer my support in this venture. Put me on the list, I will help in anyway possible."[4] An opponent countered: "I am offended at the committee's approach to the public. You come to us with what appears to be a politically driven, not yet carefully conceived, done deal and ask us what shape we want it to take. How condescending! How about asking whether we think it's a good idea or not and why?"[5]

Despite the varying individual opinions on the matter, research on industry and education in Nevada supported the idea of creating a new college. A report prepared for the state of Nevada, called *A Technology Strategy for Nevada*, concluded that, when compared with other Western peer states, and often when compared with all the states in the United States, Nevada ranked at or near the bottom on key technology indices. These included industry R&D as a percentage of gross state product, scientists and engineers per thousand employed persons, and the number of students in higher education per thousand residents. When educational attainment was defined as the percentage of residents who have at least some college experience, Nevada was last among its peer states.[6]

In 2000, the National Center for Public Policy and Higher Education published the report *Measuring Up, 2000: The State-by-State Report Card for Higher Education*. The board of the National Center and the advisory panel for the report included a virtual who's who of national higher education and public policy leaders. Drawing upon rigorous empirical data, the authors presented a state-by-state comparative profile focused on five areas. Below are the questions they asked and Nevada's "grade" in each area:

1. Preparation: "How adequately are students in each state being prepared for education and training beyond high school?" Nevada received a D+.
2. Participation: "Do state residents have sufficient opportunities to enroll in education and training beyond high school?" Nevada received a D+.

3. Affordability: "How affordable is higher education for students and their families?" Nevada received a B.
4. Completion: "Do students make progress toward and complete their certificates and degrees in a timely manner?" Nevada received an F.
5. Benefits: "What benefits does the state receive as a result of having a highly educated population?" Nevada received a C–.[7]

The Founding President

In late 1999, the Nevada Board of Regents, which oversees Nevada higher education, appointed Richard Moore founding president of Nevada State College at Henderson. To an action-oriented educator, few events could be as exciting as being part of the launching of a new college. As founding president, Moore envisioned a premier undergraduate college focused on the preparation of teachers and nurses in the area.

The former president of the Community College of Southern Nevada (CCSN), Moore had led that institution through a period of great growth and development. Moore "walked the talk." At CCSN, he had introduced innovations that helped increase enrollments, particularly of minority students and white students from poverty; improved graduation rates; and provided students with high-tech skills. High school students could acquire community college credits while they were juniors and seniors. CCSN technology centers opened on high school campuses. Having identified the absolute worst students, CCSN offered those students jobs at the community college, where they were mentored by college technical workers. Over 85 percent of those "worst" students then graduated from high school. An article in the respected *Chronicle of Higher Education* highlighted those changes.

> Under the leadership of a hard-charging president, the Community College of Southern Nevada has been transformed from what many once considered an academic dead end into a major force in [Las Vegas's] future. . . . That effort has already started with 75 students at a new high school on the college's main campus. "When you look at all the growth," Mr. Moore says, "the simplistic solution is to just have more of what you've got. The far more interesting question is, 'In that creative process of building, could you create something that could be a role model for the nation?' " While educators often talk about these kinds of partnerships, the college is unique in having taken fast, strong steps to achieve its goals, observers say. Robert S. Peterkin, an expert on urban education at Harvard University's Graduate School of Education,

believes other states may build on the models here if they save Nevada money and improve the college-going rate.[8]

During his earlier 20-year term as president of Santa Monica College (SMC), Moore concluded that the goal of preparing community college students to transfer to four-year institutions was mostly rhetoric; few students from any community college actually enrolled in a university, and even fewer graduated. He was determined to convert this rhetoric from a sham into a reality, and he did just that. Presently 1 out of every 13 students at the huge UCLA campus has attended SMC. In fact, the SMC catalog has a picture of UCLA on the cover. To accomplish this goal, Moore identified and directly confronted the barriers to successful transfer to the university. For example, he established a child care center on the campus to make it easier for single mothers with limited resources to attend classes.

Academic Planning: More Than Just a Mission Statement

In March of 2000, I met with President Moore and Chancellor Jane Nichols to draft a mission statement for the college, which was to be presented to the board of regents at its next meeting. This was an exciting, delightful meeting and afternoon. I really had the feeling that we were creating something, that the words we produced would make a difference. (My hunch was reinforced a year later, when a senior humanities professor who was considering joining the faculty told me that he had been motivated to work at Nevada State College after reading the unique, extraordinary mission statement on the college Web site.)

In advance of our meeting, we reviewed the mission statements of some neighboring colleges, including recently founded innovative institutions such as Evergreen State College (Washington), whose mission statement reads: "Teaching is the central focus of work of the faculty at both the undergraduate and graduate levels. Supporting student learning engages everyone at Evergreen—faculty and staff. . . . Students are taught to be aware of what they know, how they learn and how to apply what they know; this allows them to be responsible for their own education, both at college and throughout their lives. . . . College offerings involve active participation in learning, rather than passive reception of information." Likewise, California State University–Monterey Bay states that its campus "will be distinctive in serving the diverse people of California, especially the working class and historically undereducated and low-income populations."

Each of us brought special ideas to the drafting of the mission statement. We agreed that this institution was to be excellent and that quality was to be defined by the curriculum and instruction quality, not by institutional resources like the number of books in the library or by freshman class SAT scores. We shared a commitment to making the college accessible to poor and disadvantaged students. President Moore wanted the mission statement to include a description of the physical environment and how that would enhance learning. I wanted to emphasize value-added notions and the opportunities this college would provide to the disadvantaged and the disenfranchised in southern Nevada. Chancellor Nichols noted that the sad reality about many colleges and universities is that it takes students far more than four years to complete a four-year program, often because courses aren't available when they should be. She insisted that the mission statement include a commitment to "time-certain degree completion." At the end of the day, we had produced a mission statement which President Moore subsequently presented to the board of regents.

That summer there was a barrage of critical newspaper articles. At that time, Mike O'Callaghan, executive editor of the *Las Vegas Sun*, also wrote a regular column. A war hero who had lost a leg in Korea, O'Callaghan had served as Nevada governor from 1971 to 1979. He had pushed for both fair housing and environment protection. Mike O'Callaghan wrote: "Moore has to understand that anybody who makes greater progress than past administrators becomes the enemy of those who were happy with mediocrity. From the first day I met Moore, it was obvious to me that he was going to move too fast and be too successful for more than a few of our local education critics."[9]

At this point I thought we needed a dramatic positive event to counter the wave of negative publicity the new college was receiving. What if we were to invite the board of regents and other key figures in the planning effort for Nevada State College to meet with Barry Munitz, president and CEO of the J. Paul Getty Trust, and seek his wisdom about the political and academic issues confronting us? Prior to joining the Getty, Munitz had been the chancellor of the California State University System, the largest public higher education system in the world. He had helped found three institutions in this system: CSU–Monterey Bay, CSU–San Marcos, and the nascent CSU–Channel Islands. Dr. Munitz agreed to meet with the group from Nevada.

We met with Dr. Munitz at the Getty Museum in Los Angeles in October. Most of the Nevada Board of Regents flew to Los Angeles for this meeting, along with the few members of the college staff, led by Richard Moore. Dr. Munitz spoke about starting new colleges; the pressing issues in American higher education;

the unique social, political, and economic environment in Nevada; and higher education leadership. His presentation, done without notes of any kind, was a tour de force which engaged, informed, and energized the board of regents.

The chair of the board of regents, Thalia Dondero, asked how to settle debates about the architectural design of the campus. Munitz suggested that the state host an architectural competition. Enthused about this idea and, in fact, energized by the entire meeting, she quickly appointed a committee of educational, business, and governmental leaders in Nevada to plan and implement an architectural competition. In addition, she appointed a similar commission to review the financial plans for the new college and a third committee to develop an academic plan for the institution.

Munitz had given us some great ideas for anticipating potential issues of contention and for addressing them effectively. The most challenging academic planning task ahead of us was to decide what the students would be required to learn during their first two undergraduate years.

General Education: What Should Students Learn in College?

> My experience at Grinnell combined with other experiences, especially in the early 1940s at the University of Iowa, led me to the view that searching for the perfect liberal arts curriculum is a quest for the Holy Grail. It is never attainable but the search itself is exhilarating. If the search stops, liberal education degenerates.
>
> Howard Bowen, *Academic Recollections*[10]

Undergraduates take required general core courses, in subjects such as mathematics and foreign languages, in their first two years and then focus on their major. During the first two years, institutions require all students to take courses that will define the core of their undergraduate education. For decades, debates have raged about such requirements, prompting questions like these:

- Should general education emphasize the classics of Western civilization, or the great works from Eastern cultures, developing nations, and indigenous peoples? Does contemporary thought have a place in the core curriculum?
- Should general education courses focus on content (names, dates, and events) or on the development of analytical reasoning and critical thinking?

- Should general education courses include classes in technology that could lead to actual employment and future income?

A study by the National Endowment for the Humanities in the late 1980s profiled the decline in general education. The results of their survey "showed it was possible to graduate from 78 percent of the nation's colleges and universities without studying the history of Western civilization; 37 percent without studying any history at all; 45 percent without studying American or English literature; 41 percent without studying mathematics; 77 percent without studying a foreign language; and 33 percent without studying natural and physical sciences."[11]

A reexamination of general education requirements at colleges and universities across the nation began in the 1970s with reflection and debate at Harvard College about the undergraduate curriculum. That effort was led by the revered Harvard economist Henry Rosovsky. Rosovsky served as dean of Harvard College under President Derek Bok, and wrote one of the most entertaining and informative books about the modern university, *The University: A User's Manual*. In the first decade of this century, the university once again reviewed and revised Harvard's general education requirements. In a 2007 report, the Task Force on General Education concluded: "Our proposal is consistent with past general education programs at Harvard: it prescribes a set of requirements and calls for a set of extra-departmental courses, rather than advocate that students have a free range across existing departmental offerings in the form of an open distribution system. Since 1945, our Faculty has believed in the importance of taking a stand on the question of what students need to learn. General education is a statement about why a liberal education matters."[12]

A 1996 report by the conservative group the National Association of Scholars (NAS) had found that during the 30-year period beginning in 1964, colleges and universities gradually reduced general education course requirements in foreign languages, science and mathematics, history, literature, and philosophy. NAS president Stephen H. Balch blamed scholars who wanted to teach only in their narrow specialization and who resisted any form of hierarchy or prioritizing for students.[13] Critics of the report agreed that a reduction in requirements occurred over a 30-year period but cited pressure to make undergraduate education more job- and career-relevant as the cause. Whatever the cause, the decline in general education graduation requirements is alarming because a general college education defines what it means to be an educated person.

David Denby, a Columbia University graduate and movie critic for the *New Yorker,* has written a provocative and insightful book about the issues in general education. For decades Columbia has required an extraordinary Western civilization course, a two-semester course that may be unmatched anywhere in the higher education firmament. (My brother Joe, a successful and articulate academic, also took that course in the 1960s, and to this day, he will tell you that some of his observations and insights still can be traced to it.) In his late forties, Denby was engaged in midlife reflection and also was reading about the culture war debates about general education. Should students be required to study only classics that could be described as books written by "dead white men"? Denby decided to confront these issues directly by reenrolling in the Western civilization course he took in the 1960s.

The book he wrote about this experience, *Great Books,* is a fascinating tour de force, an examination of the meaning of classic works for contemporary men and women. It is a highly personal, highly readable account of how classic books gave Denby insights into the meaning of his own life in a sophisticated urban intellectual environment. He reflected upon being mugged while reading Hobbes and about death while reading *King Lear.*

> I could not suppress paradoxical thoughts about the left-academic critique of Western classics and the famous hegemony of the West. The *Iliad* and the *Odyssey* were both bitterly and unresolvably split in their meanings; Aristotle severely corrected Plato; Sophocles, in *Oedipus the King,* suggested that the drive for dominance—hegemony, if you like—necessarily entailed a condition of blindness; the playwrights darkened the old violent legends. . . . What was the academic left talking about? Even in its earliest stages, the central line of this tradition offered less a triumphant code of mastery than a tormented ideal of obligation and self-knowledge.[14]

Administrators, who daily face decisions about what students should learn, must ask themselves these questions:

- What weight should college curricula give to general liberal education as opposed to technical or vocational education?
- How can large institutions be structured so as to be supportive, not dehumanizing?
- How can universities increase the options for young people, as opposed to placing them on a treadmill toward a predetermined status in society?

- Is it useful or restrictive for youngsters to be taught specific competencies?
- How and where should young people acquire sophisticated skills of intellectual inquiry?
- Does occupational training really prepare people well for the job market?

How can we structure our colleges and universities so that they educate young people both for their career roles in society and for a life which will be humane, productive, and filled with the excitement of intellectual inquiry?

Colleges and universities are notoriously reluctant to change. Franklin Rhodes, president emeritus of Cornell, speaks of the "cardinal rule of academic governance: You should never do anything for the first time."

Implementing Ideals into Practice

In January 2001 the three Nevada State College taskforces appointed by regent chair Thalia Dondero began their work. Each comprised an impressive group of southern Nevada's political, business, and intellectual leaders. Design committee co-chair and regent Howard Rosenberg said of the college, "What we're building is going to be a visual sculpture that's going to last a minimum 250 to 300 years. This is something special. It hasn't been done for a long, long time and I want us to do it right."[15]

My time was committed to the academic taskforce, which was co-chaired by Dr. Jill Derby, an intellectual leader on the board of regents, and Bonnie Bryan, a civic leader. The first meeting included opening statements, needs assessments, and articulation of goals and objectives for each of the subcommittees that Dr. Derby and Bonnie Bryan appointed. The highlight of the meeting was a presentation by Chancellor Jane Nichols in which she reviewed the empirical data supporting the desperate need for a new state college in southern Nevada.

The academic taskforce concluded that every course taught during the first two years of undergraduate study should address these areas and skills: mathematics, writing, technology, and diversity and global perspectives. The plan addressed the key issues in the general education debate as it as played out across the nation's colleges and universities:

- liberal education versus vocational training
- process versus content

- concepts across disciplines
- community service

Our learning-centered philosophy emphasized "value-added outcomes." In focusing on value-added outcomes, we built on the work of Alexander W. Astin, who was the first to argue that assessment of learning in higher education should focus on "value added," as opposed to simply measuring student knowledge in the senior year. To illustrate the concept, consider two entering college freshmen; the first tests at 90 percent in, say, history and the second tests at 60 percent. As seniors, the first student tests at 92 percent and the second at 85 percent. While the first student demonstrates a higher achievement level as a senior, the second has gained more knowledge—in other words, has greater "value added."

One of the most important missions of this new college was the preparation of teachers. Lionel "Skip" Meno and I developed recommendations for teacher education at Nevada State College. Then the dean of the School of Education at San Diego State University, Meno had been a teacher, principal, school superintendent, and commissioner of education in Texas under Governor Ann Richards. We were particularly concerned that pedagogy would be infused into each course offered at the college. We wanted to build an apprenticeship model in which, during each successive year of college, students aiming for a career in teaching were engaged at a more serious, extensive, and demanding level in fieldwork experience in the Clark County School District. We proposed an integration of pedagogical issues into virtually every undergraduate course.

We reiterated that Nevada State College would be a liberal arts college with a strong emphasis on preparing teachers. Although some students would graduate and enter other fields, many, if not most, would become teachers, at least for the first part of their careers. (Many futurists now assert that most educated young people should plan on having several careers in the course of a lifetime.) Nevada State students planning on becoming teachers would be required to have a full academic major other than education.

Teacher preparation would be integrated into the undergraduate experience in two ways. First, in virtually every course, professors would be encouraged not only to present a topic, but also to discuss the pedagogical implications of that topic. In particular, they should discuss with the students how this information should be taught and should require the students to do an assignment about that question. Most educators have learned that you must fully master a subject when you are required to teach it to someone else. We hoped this would be true of Nevada State undergraduates completing these assignments. Second, starting in the first year, undergraduate students would participate in field experiences in the Clark County School District. Students who did not plan to become teachers would have field experiences at other organizations relevant to their majors.

Close cooperation with the Clark County School District could lead to a "win-win" situation, whereby the county would help develop a cadre of well-educated beginning teachers who might be more likely to stay in the profession—and in the school district—than other new teachers. We would ask Clark County to assume major responsibility for supervising these field experiences and identifying mentor teachers who could work with the students. These master teachers could officially become adjunct professors, or "faculty associates," at the college. They would have substantive input to the teacher preparation program and to the college. Not only would the senior school district officials work closely with the college faculty, but the future teachers would have a thorough general and liberal arts education to bring to their schools and to their communities. The preparation of teachers would not be an isolated instance of vocational training but would be strongly linked to general education.

Following the meeting of the board of regents at which the reports of the three taskforces were approved, *The Regent's Review* included a special progress report on Nevada State College at Henderson. The *Review* quoted Dr. Jill Derby, co-chair of the Academic and Student Services Task Force: "We were able to bring forth distinguished members of the community as well as distinguished national consultants and utilize their talents in a common process. This process has been a remarkable one and I've always believed that a good process brings good results."[16]

Political Barriers

The last week of the legislative session was the first week in June. This week, which started quietly enough, was to be the most eventful week in the history of Nevada State College.

Monday. The last scheduled day of the legislative session. For four months, both houses focused on political battles about reapportionment of districts based on the 2000 census. The legislators kept the session open for an extra 12 hours while they frantically tried to pass bills that had not yet been addressed. One of those was the bill authorizing funds for fiscal year 2002 for the establishment of Nevada State College.

Tuesday. The legislature concluded its business without addressing Nevada State College funding. Legislative leaders announced that they had not been able to find the bill during the last frantic hours and were unable to act! The college, which had received legislative approval for substantial funding in year 2 (fiscal year 2003), was left with zero dollars for the upcoming, planning and gearing-up year. This created a crisis of severe proportions, since everybody, including the legislators, expected and demanded that the college would open in September 2002. A total lack of funds for the final planning year, which would include developing the curriculum, defining majors, recruiting administrators and faculty, hiring the initial faculty, orienting the faculty, and recruiting students, placed us in a precarious situation. Amid chaos in the last minutes of the 2001 legislature, money to get the state college up and running was either forgotten or passed over for lack of time, according to university officials. The *Las Vegas Sun* reported:

> "To the best of my knowledge the bill did not pass," Chancellor Jane Nichols said this morning. "It was crazy down there last night."
>
> Legislators were expected to approve $1 million in startup money before the midnight deadline for adjournment today, but failed to do so, Nichols said. A hold up of that money could delay the opening of the college.[17]

It has always been unclear to me why the bill was not funded. Perhaps it was true that the bill simply got lost.

Wednesday. The college's vice president for planning announced his retirement from the University and Community College System of Nevada and from his current post. There was no budget, and there was no anticipated funding for his position as of July 1, just two days away. By resigning on Wednesday, he remained eligible for retirement benefits that he would have lost had he had delayed this decision.

Thursday. The board of regents met to discuss the Nevada State College funding situation. They had access to funds through a complex bureaucratic structure involving estate tax funds. However, they refused to allocate any funds for Nevada State College, and insisted that President Moore raise private funds for the forthcoming year (which was only days away). The regents repeatedly reminded Moore of the early assertions by the Henderson political leadership that they easily could raise millions of dollars in private funds to support this college.

> The Board of Regents on Friday gave Richard Moore the ultimate ultimatum: Find $1 million in private donations for the Nevada State College at Henderson or lose his job as the school's president.

Regents backed Chancellor Jane Nichols' recommendation that Moore find the money the college needs for startup costs.

"It has to be dependent on the community to support this college," she said. "It is time for a private-public partnership to work or not. We do not have money or positions to devote to Nevada State College."[18]

In four days, the legislature had declined to fund the upcoming year for Nevada State College, the board of regents had declined to allocate any funds to the college in this crisis, and the college vice president had resigned and retired. President Moore was expected by every constituency to open the doors of the college in September 2002, but there were no funds for the period July 1, 2001, to June 30, 2002. The leader of a public institution, the first college in a projected state college system, was told that his only recourse was to raise private funding for the start-up year of the college.

Former Governor O'Callaghan raged:

During the 45 years I've been in Nevada no single person has done more to promote higher education than Dr. Richard Moore. Not only did he bring our Community College system in Clark County to life, he also identified the need for a state college. His foresight and energy has been a plus for Nevadans of today and tomorrow.

Moore's dedication to promoting and improving higher education is almost beyond belief. A lesser man would have sung Johnny Paycheck's song "Take This Job and Shove It" to the regents and Chancellor Jane Nichols when hearing he wouldn't have a job as president of Nevada State College at Henderson if he didn't raise more private money. It wouldn't be difficult for a person with his talent to find an even higher-paying job. Despite the roadblocks put in his way he plods forward to make this a better place to live.[19]

Rebound and Recovery

President Moore called a meeting of advisors and supporters for October 5, 2001. In the wake of the September 11 tragedy, I chose to drive from Los Angeles to that meeting. While he was upbeat and enthusiastic in leading our discussion, it was clear to everyone that the sword of Damocles was hanging over the college. Moore distributed and discussed a short application form he had developed for use in recruiting, assessing, and admitting the college's first students. As far as I can tell, he and his assistant alone had decided that the nickname and mascot for the new college would be the Mustangs. He distributed pencils embossed with Nevada State College Mustangs and a profile of a mustang. A week later, some supporters of the college pointed out that the Mustang mascot evoked images of the most famous brothel in Nevada, the Mustang Ranch. Soon after, the mascot vanished. Everything seemed to be going badly.

In October and November, Bill Martin, the new chair of the board of regents, cajoled, pushed, and pressured some of his affluent friends to contribute to the college. He also contributed a significant amount himself. When he was finished, there were enough funds in the coffers to guarantee that Nevada State College could survive that 12-month period. After that, of course, the funding that the legislature had approved for year 2 would kick in. The crisis was over.

That fall was a disturbing and threatening time for the state economy and the citizens of Nevada. There is only one industry in Nevada: gambling. Most people who gamble in Las Vegas, Reno, and elsewhere in the state travel to Nevada by air. After the tragic events of September 11, 2001, air travel dropped dramatically, and Nevada's economy took a major hit. Thousands of Las Vegas service workers were laid off. The governor gave a televised speech in which he discussed how tough Nevada citizens were and how the state would ride out this crisis. But the events of September 11 had revealed how vulnerable this one-industry state was.

In December, I attended a meeting of consultants at which candidates for various administrative positions at the college were interviewed. The mood was upbeat. The consultants were experienced, skilled, and enthusiastic about the institution. Most of the applicants were impressive, including a woman who applied to head the nascent nursing program and a young Ph.D. psychologist who applied to become a faculty member. Things were looking up.

At this point, the entire staff of Nevada State College consisted of a president, Richard Moore, and his administrative assistant. The institution was working with a minimum budget until July 1, 2002, which was barely enough to support planning and early implementation activities. In a careful and, I thought, judicious decision, Moore decided that each consultant should be paid $1,000 per day. This was less than the usual daily fee of some of the consultants, but no one complained. We rarely had the opportunity to participate in a project with such potential.

The search to fill desperately needed positions began in earnest, with the aim of completing all new hires by the end of March. We worked closely with administrators within the Nevada higher education system office to ensure that all appropriate procedures, especially equal opportunity procedures, were implemented appropriately. We worked closely with the representatives of the accreditation sponsoring institution, the University of Nevada at Reno. Resumes flooded in for each position. My colleagues and I reviewed them and created a short list of the most promising candidates. For several of the searches, we conducted telephone interviews. Most applicants had impressive backgrounds and credentials. Most were excited about the prospect of playing a key role in launching a new institution. After the telephone interviews, we whittled each list down to a small group of finalists and invited them to Las Vegas for interviews. Richard Moore joined some of the interviews. He suggested an excellent question for the potential math professors: "Please explain why it is impossible to divide by zero." Two high-profile professors at the University of Nevada–Las Vegas were considering transferring to the new institution "down the road." The excitement and momentum were palpable.

A New Crisis

Despite our enthusiasm, the critics had not vanished. A reporter from the *Las Vegas Review-Journal* had interviewed Richard Moore several times about the consulting contracts. The legislature and other constituencies would be greatly disappointed if the college did not open its doors in September 2002. But no public funds had been authorized for the 12-month fiscal year from July 2001 to June 2002. The regents had insisted that Moore raise private funds to cover that period. Now the opponents of the college were raising questions about how those funds were being spent. Moore was asked to submit copies of all of his consulting contracts to the newspaper, an allowable request under Nevada state law. The newspaper ran articles criticizing the consulting contracts, claiming that $1,000 per day was too much to pay consultants. The chorus was then taken up by the three regents who had consistently opposed Nevada State College. Moore responded to the newspaper articles by pointing out that he had hired consultants with national reputations. Despite my experience, street smarts, and cynicism, I thought no one would take their criticisms seriously. I couldn't have been more wrong.

One Saturday in February, I received a voice mail message from Moore asking me to call him about a matter of "some urgency." When I phoned, to my complete shock and amazement, he told me that he had had lunch with the chancellor and the chair of the board of regents and had submitted his resignation as president of Nevada State College.

The next day, I interviewed the finalists for dean of Arts and Sciences. One had flown to Las Vegas from Arkansas. Arriving quite early at the University and Community College System headquarters for his interview, he went across the street to a small restaurant to have breakfast. He picked up a copy of the local newspaper and, to his shock, saw a front page story, above the fold, announcing that the president of Nevada State College had resigned and the future of the institution was in doubt. By the time he had appeared for his interview, he was understandably a little confused.

With assurances from Dr. Nichols and board chair Thalia Dondero that the college still would open in September 2002 and that we still needed a dean of Arts and Sciences, interviews for the position continued. At the end of the process, the search committee unanimously voted to hire Lois Becker. Below is an excerpt from "My Teaching Philosophy, or What Students Have Taught Me," which Lois Becker submitted as part of her application to become dean of Arts and Sciences.

> I believe in the life of the mind. I love my family. I respect physical prowess. I cherish social commitment. However, the greatest satisfaction I find is in the challenge of exercising the mind. My grandparents impressed on me the importance of education by saying: "As the Cossacks chase you from town to town, the one thing they cannot take away from you is your education." I have taken this to heart. Books are my best friends. Puzzles are my constant companions. Life is a series of intellectual adventures. I want to share this feeling with students who already know it and bestow it as a gift on those who don't. The highest goal of the university is to teach all students to think clearly, to make relevant and valid judgments, to discriminate among values, and to communicate freely with others in the realm of ideas.

From that point, events unfolded rapidly. The founding president of the institution, a vigorous creative visionary, had found it necessary to resign in the face of attacks that I had totally underestimated. The situation was summed up well in a column by former governor O'Callahan:

> Few people love their profession enough to step down from a position of power to save the goals they were attempting to reach. That's exactly what Nevada State College President Richard Moore did this week. After being battered from all sides for more than a year, Moore determined that he, not the college, was the target of the naysayers. So he stepped down and offered to return to the classroom.
>
> I'm not sure that this willingness to give up his position will stop attempts to derail the college. What has happened is the college opponents have found him to be an easy target to accomplish their ends. Moore, who thinks faster

> than most people and is unafraid to act on ideas he believes are good for education, will soon be replaced by someone or something as their new target.
>
> Moore was on the move, and so was access to higher education in Nevada. Was he moving too fast? For some people he was way too fast and he refused to be placed in the box they have built for their view of education. Trouble was brewing and both the state college and Moore became easy targets for their enemies. Because he was willing to take chances, and work outside of their box, they chose him as the target of opportunity.
>
> On a level playing field Moore would still be building the state college. He has learned that politics never provide a smooth road to success in achieving worthwhile goals. Now he will have an opportunity to see somebody else reach those goals. The foundation for educational progress in the form of a state college has already been built. Eventually it will become a reality no matter how much a few people attempt to keep it from helping more young people realize their dreams and fulfill the needs of Southern Nevada.[20]

To a Las Vegas resident who regularly read the newspaper yet who did not know the facts about what was happening at Nevada State College, it must have seemed that the new institution was illegitimate; had been developed by politicians, not educators; and was being led and designed by people who couldn't manage, who couldn't control the school's finances, and who couldn't be trusted. A steady drumbeat of newspaper articles fueled by the opposing regents and others who resented the idea of a new institution created mirage crises out of trivial events. Dozens of articles quoted regents and others who felt the consultants were being paid excessively. (One consultant routinely submitted invoices on which he listed his usual consulting fee of $3,000 per day and then explicitly listed a "discount" of $2,000, yielding a net fee for his work in Nevada of $1,000 a day. He did this a year before the controversy over consulting fees erupted. He was enthusiastic about his work in Nevada but wanted to indicate for the record that he was being paid less than his usual consulting fee.) Of course, the final irony was that the president and staff of the fledgling college were being scrutinized according to state fiscal policies after having been told that they would receive no state funds, but must raise private money to support the operations of the college.

Epilogue: Success

By spring 2004, enrollment at Nevada State College was healthy and growing, including some junior and senior transfer students. Thirteen seniors completed their baccalaureate requirements and graduated from the Nevada State College in May. I received an official invitation and attended the event.

The graduation was well planned and inspiring. Virtually all the major characters in the struggle to launch the new college attended. Richard Moore attended as a member of the economics faculty. The new president, Dr. Kerry Romsburg, went out of his way to introduce Moore as the founding president. A short film profiled each of the graduates. I was struck by the limited economic means they brought to their college education and by what each of them must have sacrificed to make their college dream come true. The most moving moment in the ceremony occurred when a spokesperson for the graduating seniors presented President Romsburg with a $1,500 contribution from the graduates. Given the collective net worth of the graduates, this was a truly generous gift.

Several days after the graduation, Romsburg announced suddenly that he was leaving the presidency of Nevada State College to become president of a college in Florida. Following an interim president, it was announced that Dr. Fred Maryanski, a computer specialist who was associate provost of the University of Connecticut, had been selected as president of Nevada State College. At that point, the university had been in operation for 27 months, and already it had had five presidents.[21] Today the college is a solid bustling enterprise, with an enrollment of more than 2,000 students.

Nevada State College would not have survived had Richard Moore not been the founding president and had he not resigned when he did. Study the management and leadership literature. It is rare, if not unprecedented, to find a leader who relinquishes his position so that the organization might achieve its mission.

CHAPTER 8

The Value of a College Education in the Global Economy

> When the capital development of a country becomes the by-product of the activities of a casino, the job is likely to be ill-done.
>
> John Maynard Keynes

In the introduction I mentioned the competition between contestants with "book smarts" versus those with "street smarts" on Donald Trump's blockbuster television program *The Apprentice.* This contest pitted a team of college graduates against a team of high school graduates. The show prompted me to write an opinion piece in which I questioned what people can learn from an admittedly entertaining but superficial television show.[1] In the article I also made the major arguments in favor of a college education. It ran on the Knight-Ridder newswire service and appeared in eight or nine newspapers. I received some very interesting responses to the article. This piece seemed to touch a nerve with many people.

Lee Halyard, a Virginia real estate expert, wrote that he had never attended college but was enormously successful in the business world. He earned an impressive income in real estate, served as an expert witness in court cases about real estate, and taught a "Principles in Real Estate" licensing course. Still, he felt much was missing from his life because he had not been to college.

> I read with great interest your op/ed which appeared in our local paper in Chesapeake Virginia. I am one of the Street Smart folks.
>
> I own my own company.
> I teach principles in Real Estate for State Licensing.
> I am an expert witness for eminent domain cases here locally.
> I am living a fantasy I would've never thought possible.
>
> However . . . over the past few years while I routinely have hauled down [a good income] and acquired significant assets, I'm becoming more and more

> cognizant that I lack "Broad Based Knowledge," knowledge I would've acquired and owned and have been able to utilize daily had I been to college. This is the point that is sorely missed by this show. I had to earn a large income to learn a large income wasn't everything.
>
> When it comes to buying or selling real estate, I'm probably one of the better informed folks around town here in that regard, however should I find myself in a debate with regard to e.g. history, politics, even when I speak to other business owners about running a business, I'm becoming ever more increasingly aware, that I don't posses some of the knowledge they do, that they probably acquired in college.
>
> I'm 45 years old with a company to run and responsibilities galore so starting from scratch at a 4 year school isn't an option, but its something I think about from time to time, and something my two daughters won't miss out on.[2]

In a separate e-mail message Halyard wrote: "Please feel free to quote me in any area you see fit. If it'll help one of the kids stay in college, and to not just try to run down some dream. If you were here or anywhere nearby I'd be pleased to come and do it in person. I now feel it's just that important! You can call me and put it on a speaker phone and I'll tell your whole class!"

In our exchange, I mentioned that it was not too late for him to earn a college degree—or, for that matter, a graduate or law degree. My mother returned to her undergraduate studies after raising three children and graduated from Skidmore College in her 40s. I was proud to attend her graduation. She subsequently earned a master's degree in social work and worked for many years as a psychiatric social worker.

There are a number of students who know what their career goals are—and those careers do not always involve college. Not everybody should be encouraged or forced to attend college. However, it is not true that some students are not smart enough for college. For years the focus of my work has been research that reveals that virtually every student has the capacity to achieve in high school and in college. I don't believe that every young person needs to attend college. I do believe, however, that for the vast majority of students, a college education will improve career prospects, increase lifetime earnings substantially, and enrich lives immeasurably. Students who wish to pursue careers in STEM, beyond work as technicians, surely need a college education.

To bring this entire debate full circle, back to Nevada, Libby Parker, who worked for one of the newspapers in Las Vegas, wrote the following in response to my op-ed piece about Donald Trump and the value of college:

> Living in this fine city and trying to convince the general public that a college education is, in fact, worth something is incredibly frustrating.
>
> I grew up here and am a product of our public school system. I went on to attend UC Berkeley as a Regent's & Chancellor's Scholar, where I received my degree in Sociology. Having held 3 jobs here since college graduation, I can attest to the fact that employers here are not entirely impressed by a college degree (or for that matter, where you earned one!). This flies in the face of the "Best Public University in the World" attitude that encompassed my life at Cal!
>
> And I believe this is where the problem lies. In our money-hungry, "gambling is king" economy, the value of an education is not realized. Employers need to be aware of this and change their attitudes and their actions. I cannot tell you how many job postings I've seen that, in any other city would probably require a college degree, but definitely do not set that tone here.
>
> Worse yet, many girls I knew in high school went on to become strippers and cocktail waitresses and never gave a second thought to higher education because, hey, they can make 6 figures taking off their clothes!
>
> So, now that I've shared with you my frustrations about what's wrong with this town, thank you for your column—I've cut it out and will be sharing it with as many people as possible.[3]

The Lifetime Financial Benefits of College

As we saw in chapter 7, the new four-year college in southern Nevada was a test case for addressing the value of a college education. In an effort to underscore the lessons from that debate, I present earnings data from that period (2000–2002) as well as more recent data.

In the early part of the decade, the Nevada economy was experiencing exhilarating growth. Then Nevada was hit hard by the economic impact of the 9/11 attacks. Following that downturn, the state recovered, and the boom period continued. However, there were signs that the state's economy was vulnerable. The nationwide crash and recession that began in September 2008 devastated the Nevada economy. From October 2008 to October 2009, gaming revenues in Nevada dropped 11.6 percent.[4]

Table 8.1 shows the median weekly earnings in 2000 and 2009 for U.S. wage earners 25 years and older with differing levels of education, as reported by the Bureau of Labor Statistics. In 2009, the earnings for workers with at least a bachelor's degree are significantly more than for those without a degree.

TABLE 8.1

Correspondence between education level and median weekly earnings for U.S. wage earners 25 years and older, 2000 and 2009

Education	2000 earnings	2009 earnings
Less than a high school diploma	$360	$454
High school diploma or equivalent	506	626
Some college	598	699
Bachelor's degree	896	1,025

Source: U.S. Department of Labor, Bureau of Labor Statistics, "Working in the 21st Century," www.bls.gov/opub/working/data/chart6.txt (year 2000), and "Employment Projections," www.bls.gov/emp/ep_chart_001.htm (year 2009).

While earnings increased in proportion to level of education attained, unemployment rates decreased as education levels increased. The reverse was also true: unemployment rates increased for those with less education. In 2000, the unemployment rate for individuals with only a high school degree was 3.5 percent. Nine years later, that figure nearly tripled, to 9.7 percent.[5]

The Center on Educational Policy in Washington, DC, makes this point in a guide for middle and high school students on the financial benefits of college, how to better prepare for college-level work, and how to improve the chances of being admitted: "The income gap between people with no college education and people with some college education has widened in years—mostly because wages for workers without college are going down."[6]

Not only are young people from poor families less likely to go to college, they're more likely to drop out of college. According to the Census Bureau, almost one in three people now in their mid-20s started college but did not receive their bachelor's degree. The *New York Times* reported in 2005 that "on campuses that enroll poorer students, graduation rates are often low. And at institutions where nearly everyone graduates—small colleges like Colgate, major state institutions like the University of Colorado and elite private universities like Stanford—more students today come from the top of the nation's income ladder than they did two decades ago."[7]

Economist Gary Becker reports that the financial benefits of attending college continue to increase: prior to the 1960s, college graduates earned about 45 percent more than high school graduates. By the 1980s the disparity in earnings rose to 65 percent.[8] The earnings gap may have contracted slightly during the current recession, but the lesson from these data is clear: in general, there is an economic payoff from a college degree. The linkage is not perfect. Some college graduates have not been able to find jobs during this recession. And some technicians who did not attend college are currently in demand. But the general economic benefits of a college education, particularly one in which some marketable skills are acquired, seem clear.

Of course, a full understanding of the financial benefits of going to college must also consider the costs incurred in college as well as the opportunity costs represented by the income that would have been earned while working instead of going to college. But the increased average income of college graduates quickly erases those costs in a matter of a few years. Furthermore, future income is associated with a number of factors, including the quality of the institution. The author of a 1998 paper concluded: "All things being equal, if one desired to maximize post-graduate earnings, she would choose a high quality college or university, major in a lucrative area such as health or engineering, and strive to attain a high grade point average over the course of her studies."[9]

The decision in 1999 to build a new college in Henderson, Nevada, was made partially to improve the earnings and career prospects of students in the area. The U.S. Census Bureau prepared work-life estimates of lifetime earnings for workers aged 25–64 with differing levels of education drawing on data from the Current Population Survey (CPS) about 1997–99 earnings. Table 8.2 shows the earnings of high school dropouts, high school graduates, and college graduates on an annual basis. As this table makes clear, for the time period shown the annual income difference between full-time workers with bachelor's degrees and full-time workers with high school diplomas was $21,800. Over a 40-year career, that difference grows to $872,000.

Yet it is hard to convince wage earners in Las Vegas that a college education is worthwhile. A skeptic could easily argue that anyone who can get a high-paying job in Las Vegas doesn't need a college education. I would argue the opposite point of view. First, a college education enriches your life and gives you the tools for a better life, in which you can enjoy living beyond the monetary return on investment. Second, even if one could remain a parking attendant earning good money for decades, the work itself is not intrinsically rewarding.

TABLE 8.2
Average annual earnings of workers ages 25–64 by educational attainment level, 1997–1999

Education	All workers	Full-time, year-round workers
Less than a high school diploma	$18,900	$23,400
High school diploma	25,900	30,400
College degree	45,400	52,200

Source: U.S. Census Bureau, *The Big Payoff: Educational Attainment and Synthetic Estimates of Work-Life Earnings*, U.S. Department of Commerce brochure, July 2002, p. 2.

Parking cars will become boring after a few years, if not a few months. Third, service jobs depend on a vibrant economy. In recessionary times like these, hospitality jobs may not be as secure as they once were.

The Nonmonetary Value of College

What is the full value of a college education? College affords young people the opportunity to grow and be challenged in a variety of ways. Put differently, one cannot measure the value of a college education solely by comparing investment dollars (for tuition, books, room and board, etc.) with postgraduate earnings. For one thing, the competencies and intellectual skills learned in college, while not strictly job related, affect students over the course of their careers. Second, developmental changes enrich the leisure time of the former students as adults. There is some evidence that job satisfaction and leisure-time satisfaction are correlated. Developmental changes and learning in college will enrich both the lives and the subsequent careers of today's students.

A debate has been conducted in the professional and popular literature about the impact of college on subsequent career development and earnings. Some critics have argued that the investment by a student and his or her parents in college does not pay off. I question whether the *investment* in college by a student and his parents can be measured only in dollars and cents, and whether the *effects* of college can be measured only with that criterion. Both the investment and the payoff from college involve many dimensions of the human experience. Some of the payoffs relate to subsequent work experience; some do not.

The true value of college is more than just economic gain. Even the relationships between education and work extend beyond economic benefits to include how knowledge gained in college is or is not applied to careers; how the college experience can prepare students to adapt to the rapidly changing world of work; and how a liberal education—in the classics, for example—prepares a person for a more rewarding life both in and out of work. One goal of higher education is realizing the full potential of individuals and of society. This goal corresponds closely to the goals of human life. As Alexander Heard, then chancellor of Vanderbilt University, remarked, "Our largest common goal in higher education, indeed in all education, is to create and stimulate the kind of learning that breeds strength and humor and hope within a person, and that helps build a society outside him that stirs his pride and commands his affection."[10]

Without college, the poor children of Las Vegas are likely to become poor adults. A few of them may become compulsive gamblers, with desperate financial needs.

Besides the old charges of usury and fencing, Las Vegas pawn shops have another image they would like to disown; they are often seen as overpriced banks for gamblers. Some pawnbrokers insist that gamblers make up only a small percentage of their customers. Location is obviously relevant here; Kronberg guesses that up to 50 percent of his customers pawn possessions in order to continue gambling. Players can make last-chance bets by giving up valuables, from a watch or a wedding ring to gold teeth and even cars. "Need Money Fast? Call Auto Pawn." Parachutes and rodeo saddles are among the items Las Vegas pawnbrokers recall crossing their counters. At Bargain Pawn, two human skulls (presumably from a medical school, though one never knows) are proudly displayed by owner George Bramlett as "trophies" in a glass case in his office.[11]

Nevada's Economic Future

The failure of Las Vegas and southern Nevada to plan to conserve and develop human resources is paralleled by the failure of Las Vegas to conserve and develop water resources. As I write these words, Las Vegas is suffering through a terrible drought, a drought that threatens to slow the unbridled population and economic growth of the past few decades. "If they don't take steps immediately to curb rampant growth, they're jeopardizing the future of the entire community," says Larry Paulson, scientific adviser to the nonprofit Nevada Seniors Coalition. "One of these days, we have to come to terms with the fact that we live in a desert, that we can't continue to have unlimited growth on a limited water supply. . . . It illuminates the neon city of Las Vegas, whose annual income is one-fourth the entire gross national product of Egypt—the only other place on earth where so many people are so helplessly dependent on one river's flow."[12]

The present drought has been described as the worst since the Middle Ages. Char Miller, a history professor at Trinity University in San Antonio observed: "The broader question isn't whether limiting growth is the appropriate measure, but whether it's actually possible. . . . They don't have the political will in Las Vegas to do it."[13]

A direct correlation exists between human resources and economic issues. Unlimited economic growth cannot be dependent solely on a limited economic base—in Nevada's case, revenues from gaming. A more diversified economy and a skilled labor force are essential. But does the political will exist in Las Vegas to build the schools and colleges that will educate that skilled labor force?

Las Vegas was perhaps the hardest hit municipal economy in the financial meltdown of 2008. A *Time* magazine cover story about the city's woes labeled it "Less Vegas." But years before, the signals had been clear about the vulnerability of Las Vegas to economic hard times. In March 2006, the unemployment rate in Nevada was 4.2 percent. By December 2009, it reached 13.0 percent. The only other state with a higher unemployment rate in 2009 was Michigan, where the statistic was 14.6 percent.[14]

An October 2000 article in the *Wall Street Journal* highlighted some of the weaknesses in the Las Vegas economy. Noting that the city had the highest growth rate in employment in the nation over the previous year (6.4%), the author observed that income gains for households in Las Vegas nevertheless lagged far behind the national curve. During the previous year, household income in Las Vegas had grown 4.3 percent, as contrasted with the leading cities on that indicator: Austin, Texas, at 8.4 percent; San Jose, California, at 8.2 percent; and Seattle, Washington, at 8.1 percent.[15] Clearly, job growth was concentrated on service jobs that do not yield high incomes.

But how do you attract high-tech firms if you don't have a college-educated, skilled labor pool? How can a city provide able workers if it has too few colleges for its high school graduates?

After reviewing data, charts, and statistics about the economy of Nevada, past and present, I believe that the economic future of Nevada, especially southern Nevada, may be bleak. As we have already seen, it has only one viable industry: gaming. State political and business leaders have taken few steps to diversify the economy. To use the gambling metaphor, they are not hedging their bets. There are many indicators that gambling revenues will decrease over the next 10 or 20 years. When Nevada tries to attract other industries, the lack of a skilled labor pool will prove to be a decisive factor. I believe that gambling revenues will fall slowly but steadily, placing Nevada in permanent economic difficulty within 20 years, if not sooner.

When Las Vegas was launched as America's gambling mecca, it had no competition. Atlantic City legalized casino gambling and developed a smaller version of Las Vegas in New Jersey, but it did not pose serious competition. But, gradually and inexorably over the past 15 years, southern and midwestern states have legalized gambling, often restricted to specific locations like "riverboats" that are permanently anchored outside major cities like New Orleans and Biloxi, Mississippi.

The Nevada economy is also threatened by the rapid growth of state-supported gambling in states like Rhode Island, South Dakota, and Oregon. All but 2 of the 50 states now have some form of legalized gambling.

- In Rhode Island, gambling has become the third largest segment of the state's income, ahead of the corporate income tax.
- In Delaware, gambling revenues made it possible to reduce the top personal income tax rate from 8.4 percent to 5.9 percent, a huge reduction.
- In South Dakota, gambling revenues made it possible for the state to reduce property taxes by 20 percent.

According to David Knudson, chief of state to former South Dakota governor Bill Janklow, the reduction in property taxes "only increased our dependence on gambling . . . the biggest addict turns out to be the state government that becomes dependent on it."[16]

Most states derive less than 10 percent of their income from gambling. The most direct measure of Nevada's dependence on gambling revenues (and, therefore, its vulnerability to a downturn in those gambling revenues) is its percentage of revenue derived from gambling: a whopping 42.6 percent.

Legal casinos on Native American tribal lands are another competitive factor linked to decreasing gambling revenues for Las Vegas. There is every reason to believe these trends will continue and that Nevada gambling revenues will continue to decline.

A front-page article in the August 25, 2004, issue of the *Los Angeles Times* was headlined, "California on path to become nation's gambling capital." The author, Dan Morain, concluded: "In time, according to gambling industry officials and economists, the Golden State almost surely will pass Nevada as the nation's biggest gambling venue. The financial situation is much more complicated than simple competition, since Nevada investors, e.g., the Harrah Corporation, are investing in California tribal casinos." The article attributed the growth in gambling in California to the spread of Native American casinos.

The attempts during the past 15 years by Las Vegas to recast the city's image can be interpreted as desperate marketing moves. Largely under the leadership of Steve Wynn, Las Vegas made a major attempt to become a "Disneyland in the desert" with many attractions for families and children. The hope was that adults drawn to the family attractions and associated low hotel room rates would still drop considerable sums gambling. However, in the past few years, gaming industry leaders have concluded that this gambit did not work. Las Vegas is once again marketing its old image as Sin City, where one can do almost anything. Their famous advertising campaign slogan is "What Happens Here, Stays Here."

The terrorist attacks of 9/11 quickly and dramatically revealed the vulnerability in the Las Vegas and Nevada economies. For weeks and months after the tragedy, many Americans were afraid to travel to Las Vegas by plane. Gambling revenues dropped sharply in the fall of 2001. Casinos and hotels laid off workers. Interestingly, the economic damage was less significant in northern Nevada, where there has been a significant attempt to diversify the economy beyond gambling. A Nevada newspaper article that appeared two months before 9/11 had contained some warning signals:

> Signs of slowing are everywhere. As of June, the Southern Nevada Index of Leading Indicators had languished at around 128 for over a year. Recession worries have begun. The building boom of the 1990s has not yet been assimilated. The weakening economy in California is also worrying, since more than half of the gaming profits come from California visitors. While casino profits rise occasionally, as in May 2001, that implies slowdown as well: slot machine wagering has gone flat, "reflecting economic uncertainty that has plagued the country." Most striking, the top four "table" games—not slots—were up, baccarat most of all, 26.5%. That means that high rollers are accounting for too much of the increase, a bad signal, because the slot players fill the hotels. They are the bread and butter, the "middle market."[17]

Ill-Prepared for the Information Economy

Drive through Amsterdam, New York, a small city not far from where I grew up, and you will see once-thriving mills and businesses that are now nothing more than empty buildings with broken windowpanes. Amsterdam's fabric industry folded years ago, before moving to the South and overseas. Lowell, Massachusetts, once a hub of woolen mills, has met a similar fate. When a one-industry town, city, or region loses that industry, the economic effects are catastrophic.

The economy of Oregon traditionally has been heavily dependent upon logging and timber. Several decades ago, the processing of logs moved from Oregon sawmills to offshore facilities. The lumber companies found that it was cheaper to ship logs across the Pacific for processing in Japan and other Pacific Rim countries than it was to process them in a local sawmill. The Oregon economy has yet to recover from this loss of work. Between 1990 and 2005, Oregon has experienced scores of plant closures and the loss of more than 20,000 jobs.[18] The decline can be traced to four factors:

1. protection of some forests by environmentalists concerned about endangered wildlife like the spotted owl
2. wildfires
3. sawmill competition from foreign countries and other states
4. plant pathology . . . [as when] a fungus called Swiss needle cast sharply reduced the production of Douglas fir in Oregon[19]

In 1990 author John Mitchell visited Sweet Home, an ironically named Oregon mining town suffering severe economic decline.

> Sour times aplenty were waiting in Sweet Home when I got here. Halloween it was and no funny faces or dispensation of treats. Willamette Industries, in Portland, had just announced it would be closing its Foster stud mill the end of January, twenty-six workers out of a job. And that would come on top of seventy-four other Willamette employees out of full-time work since the previous January. Scores more than that if you took the layoffs back beyond the closure of Williamette's Midway veneer operation in April of last year to the shutting of its Sweet Home old-growth sawmill in 1989 and the demise of its Griggs plywood facility a few years before that. And forty more if you picked up Clear Lumber's pink slips as well. For Linn County as a whole, down at the mill and out in the woods, the state employment people were tallying 990 lost jobs altogether since August 1988.[20]

A statement by the Sweet Home Economic Development Group articulated a goal that could have as easily described southern Nevada a few years later: "Reduction in timber base, modernization of mills, and levels of exports overseas have caused dramatic decreases in available jobs directly and indirectly associated with the timber industry. The opportunity related to the trend is that *forced economic diversification* in the long-term will benefit the community."[21]

The Need for a Skilled Labor Pool

Businesses considering moving to the Las Vegas area, especially those in the high-tech industry, will need highly skilled, college-educated workers. Unless more colleges are built, they will find a workforce that still ranks at the bottom or close to the bottom in the nation in terms in the percentage of people who have bachelor's degrees. *It would not be an exaggeration to argue that the best move Nevada has made to protect its future economy is the establishment of Nevada State College at Henderson.*

TABLE 8.3
Metropolitan areas with the highest and the lowest "brains to brawn" ratios on William Frey's scale

City	"Brains to brawn" ratio
Highest-scoring cities	
Minneapolis–St. Paul, MN-WI	3.53
Seattle-Tacoma-Bremerton, WA	3.04
Raleigh-Durham-Chapel Hill, NC	2.66
Denver-Boulder-Greeley, CO	2.66
Washington-Baltimore, DC-MD-VA-WV	2.46
Austin–San Marcos, TX	2.41
San Francisco–Oakland–San Jose, CA	2.32
Lowest-scoring cities	
Houston-Galveston-Brazoria, TX	1.12
Greensboro–Winston-Salem–High Point, NC	1.07
New Orleans, LA	1.01
San Antonio, TX	0.99
Los Angeles–Riverside–Orange County, CA	0.90
Miami–Fort Lauderdale, FL	0.88
Las Vegas, NV-AZ	0.79

Source: William H. Frey, "Charticle," *Milken Institute Review* 4, no. 3 (2002): 4.

William H. Frey of the Milken Institute has calculated a "brains to brawn" ratio for metropolitan areas. The brains to brawn ratio is a comparison of the number of college graduates to the number of high school dropouts—in other words, the ratio of both ends of the college attendance spectrum. He notes that "during the 1990s, the number of people with a sheepskin increased by well over a third, while high school dropouts declined." He adds, "States and local communities wishing to improve their tax bases, and cash in on the knowledge economy, have done their best to lure the best and the brightest."[22] Frey's ratio is intended to yield a single number which measures how well the state and local communities have succeeded. Table 8.3 lists the metropolitan areas with the highest and the lowest brains to brawn ratios.

As can be seen in the table, Las Vegas has done a poorer job of educating and recruiting skilled workers for the knowledge economy than any other metropolitan area in the United States. A ratio below 1.0, like the Las Vegas ratio of 0.79, indicates that the number of high school dropouts exceeds the number of college graduates in that city. Frey concludes, "It seems we're getting smarter all the time, but some places are getting smarter than others."[23]

Competing in a Global, High-Tech Economy

In his book *The Work of Nations: Preparing Ourselves for 21st Century Capitalism*, Robert Reich describes the emerging global economy and how it is reshaping discussions of America's competitiveness: "The competitiveness of Americans in this global market is coming to depend not on the fortunes of any American corporation or on American industry, but on the functions that Americans perform—the value they add—within the global economy."[24] Reich argues that an economic analysis of jobs and productivity cannot be based on the traditional Census Bureau classification of jobs, a taxonomy that is inherently linked to the old industrial occupational structures. In its place, he defines and discusses three emerging categories of work. The first two are routine production services (which include most of the traditional blue-collar, assembly-line jobs as well as repetitive work in today's service and information industries) and in-person services (also routine jobs but performed on a person-to-person basis and requiring that the worker interact with others pleasantly and courteously). The third and most important category, according to Reich, includes symbolic-analytic services, which Reich describes as "all the problem solving, problem identifying, and strategic brokering activities . . . the manipulation of symbols—data, words, oral and visual representation" (177).

Symbolic analysts may be scientists and engineers, but they may also be public relations executives, investment bankers, lawyers, real estate developers, management information specialists, organizational development specialists, systems analysts, architects, writers, editors, or musicians. Reich notes that symbolic analysts "often work alone or in small teams, which may be connected to larger organizations, including world-wide webs. *Teamwork is often critical.* Since neither problems nor solutions can be defined in advance, frequent and informal conversations help ensure that insights and discoveries are put to their best uses and subjected to quick, critical evaluation" (179; emphasis added). Reich believes that despite the ills that have befallen American education, many American children—he estimates 15 to 20 percent—are being well educated for symbolic analytic work. The question, of course, is whether 15 to 20 percent is enough.

Other countries may be doing a much better job. For example, "Japan's greatest educational success has been to assure that even its slowest learners achieve a relatively high level of proficiency" (228). Consider, too, that "the software engineer from Belmont, Mass., working on a contract for Siemens, which is financed out of Tokyo, the routine coding of which will be done in Bulgaria, the hardware for which will be assembled in Mexico, is a true citizen of the global economy."[25] Again we must ask, are we doing enough, and are we doing enough for enough people?

Reich now differentiates global symbolic analysts from national symbolic analysts. The latter group "still work within a national economy . . . and are at the core of their nation's middle class." In contrast,

> most global symbolic analysts have been educated at the same elite institutions—America's Ivy League universities, Oxford, Cambridge, the London School of Economics or the University of California Berkeley. They work in similar environments—in glass-and-steel office towers in the world's largest cities, in jet planes and international-meeting resorts. And they feel as comfortable in New York, London or Geneva as they do in Hong Kong, Shanghai or Sydney. When they're not working—and they tend to work very hard—they live comfortably, and enjoy golf and first-class hotels. Their income and wealth far surpass those of national symbolic analysts.[26]

The skills required by Reich's symbolic analysts are echoed in the American Library Association's definition of information literacy:

> To be information literate, a person must be able to recognize when information is needed and have the ability to locate, evaluate and use effectively the needed information. . . . Ultimately, information-literate people are those who have learned how to learn. They know how to learn because they know how knowledge is organized, how to find information, and how to use information in such a way that others can learn from them. They are people prepared for lifelong learning, because they can always find the information needed for any task or decision at hand.[27]

The U.S. economy will require workers with increasingly greater skills. Sue Berryman, director of the National Center on Education and Employment, offers a description of the kinds of changes that have taken place within one industry.

> In the insurance industry, computerization has caused five jobs to be folded into one, known as a claims adjuster. The job occupant is less an order taker than an advisory analyst. He or she has to have good communication skills and be able to help diagnose the customer's needs through an analytic series of questions and answers. The person needs less specific and splintered knowledge—the ability to understand multiple arrays of information, the rules governing them, and the relationships between arrays. He or she also needs to be able to frame answers to less standardized requests. Insurance companies used to hire high school dropouts or graduates for the five jobs.

> They now hire individuals with at least two years of college for the restructured claims adjuster jobs.[28]

One significant blank spot in workers' skills set is a knowledge of statistics. Brian Joyner, a quality-control consultant who works with Fortune 500 companies, says that it is difficult to overestimate the problem and that "it's just as bad in the executive suites as it is on the factory floor." Every week, managers get data that fall within normal ranges of error, but because they don't understand the concept of variation, "they end up reacting to noise instead of statistically significant information."[29]

The Role of Higher Education in the Global, High-Tech Economy

> One fact that most drive-by experts never fully appreciate is that Las Vegas is the toughest factory town in America. Behind its polished marble floors and larger-than-life themes is the greatest cash-generating machine the world has ever known. Casino dealers, in their deceptively clean black-and-white uniforms and manicured nails, work the assembly line around the clock. Their toil is cleaner than the lot of coal and diamond miners, but the jobs have a lot more in common than you might think. Dealers are essential to the function of the great casino machine, but they are all but nameless, often abused, and easily replaced.
>
> H. Lee Barnes, *Dummy Up and Deal*[30]

Colleges and universities can have a profound effect on the surrounding region. Put simply, the cities, states, and regions of the United States that have thriving high-tech economies all have excellent and plentiful colleges and universities.

After World War II, Boston's economy was dying because of the decline in the local textile industry. High-tech firms, including computer software companies, eventually grew out of the synergistic efforts of area universities, including Harvard and the Massachusetts Institute of Technology, and local scientists and business leaders. The Raleigh-Durham area of North Carolina was in decline until a concerted effort was made to develop the Research Triangle—high-tech firms connected to the University of North Carolina at Chapel Hill, Duke University, and North Carolina State University. The North Carolina makeover was catalyzed and facilitated by a series of National Science Foundation Science Development grants, a program that I describe in chapter 9. Similarly, Science Development grants to both the University of Pittsburgh and Carnegie Mellon

University enabled Pittsburgh to recover from the decline of the steel industry and emerge as another high-tech center.

Colleges and universities fuel economic growth and development. In his books *The Rise of the Creative Class* and *Cities and the Creative Class,* Richard Florida has advanced the theory that cities grow and become a magnet for businesses when they are open to a diverse population. He acknowledges that universities play a role in promoting diversity, although I believe he does not give universities as much credit as they deserve. I would argue that the presence of universities is a fundamental exogenous factor that then creates a greater openness in a city for accepting people of diverse backgrounds. This acceptance, as Florida has argued, then leads to economic development. He argues that we shouldn't assess economic growth simply on the basis of population growth in a city, but rather should look at the quality of that growth, which is reflected in wages and income. He cites Las Vegas as a measure of a rapidly growing city that lacks growth in high-quality jobs: "Las Vegas—a region typically held up as fast-growing . . . but which ranks low on my indicators—is a good example of what a *low*-quality growth center looks like. It's true that between 1990 and 2000, Las Vegas ranked first in population growth and third in job growth. But in *per capita* income growth, a measure of how much people make at their jobs, it was a miserable 294th out of the 315 U.S. Metropolitan Statistical Areas that existed in 1990."[31]

It's clear that colleges benefit cities and regions by providing highly skilled workers. Likewise, a college education benefits individuals in both monetary and nonmonetary ways. Universities also benefit cities and regions by generating research projects and programs that benefit society, providing critical learning opportunities for STEM students, and attracting high-tech firms. We turn next to the importance of university research in STEM.

CHAPTER 9

Supporting University Research

> Bleibermacher has managed to parlay all of this into a remarkable reputation. Until he came to our place, he had never held an academic job. He didn't have to. Guggenheims, Sloans, Revsons, MacArthurs came his way one after the other. What they thought they were supporting God only knows. I suppose each one reasoned that the last place that had given him a grant must have known what it was doing. It reminds me of a man I know who found himself stateless after the Second World War. His wife stitched together a document that looked like a passport. He said the real problem was the first visa. Some country . . . gave him one, and after that he got visas everywhere.
>
> Jeremy Bernstein, "The Faculty Meeting"

One of America's great advantages in the global economy is our higher education system. In sharp contrast to the reputation of its elementary and secondary schools, U.S. colleges and universities are the envy of the world. To this day, American scientists and engineers set the pace for innovation. American STEM students have the valuable opportunity to learn from the world's leading researchers and to participate in research.

Most university research in the United States is supported by federal tax dollars. In financial support, as with every other aspect of STEM, rigorous evaluation and accountability are essential. Distribution of federal research funds is a central concern. The concentration of federal science funds at top-tier institutions limits the productivity of brilliant junior STEM professors at second- and third-tier universities. Furthermore, scientists at lower-prestige schools may subsequently be unable to demonstrate to undergraduates what the research process looks like. And, of course, they may be unable to engage undergraduates directly as participants in that research process. College students who hear young professors talk about the excitement of research but note that those same professors are not conducting much research are less likely to choose careers as scientists.

This situation is made more serious by the fact that there are vast numbers of students, many of them highly capable potential scientists, enrolled at these universities. While some of the recent literature about future scientists focuses on the Ivy League and elite research universities or selective liberal arts colleges like Antioch or Pomona, the fact is that there are far more students enrolled at the large state universities.

The perennial debate about the dispersal of federal science funds is largely political. Federal science funds are authorized and appropriated by congressional committees, whose members often come from states other than those that receive the lion's share of federal funding (such as Massachusetts, New York, and California). But it is not simply a political question, a case of "pork-barrel science." The legislation creating the National Science Foundation underscored, as has all policy since then, the basic funding criterion of excellence. Legislation also recognizes and endorses the notion of "geographical equity," and the NSF has created a series of programs and policies aimed at building and maintaining strong science programs at non-elite institutions across the country.

In several research projects I focused on the impacts of academic demographics on the productivity of scientists. In one study, our research team conducted field visits in each of seven states that competed for early funding in the NSF's Experimental Program to Stimulate Competitive Research (EPSCoR), a program I discuss at length below. Those seven states ranked at, or close to, the bottom in annual federal science funds awarded. The visits provided an opportunity to interview hundreds of scientists and science managers about barriers to the successful development of a research career. We also analyzed survey data retrieved from 60,000 university researchers and examined the dispersion of many of the nation's outstanding scientists, particularly young scientists, to peripheral higher education institutions as the result of a shrinking academic job market. Once on the faculty, scientists at such institutions found it virtually impossible to develop productive research careers. In the 21st century, scientific progress is greatly determined by institutional environments and is no longer strictly the result of individual creativity (if it ever was).

We drew several conclusions from our field visits to these seven states:

- A growing army of highly capable "forgotten scientists" are employed in "peripheral" U.S. universities.
- Relatively small, specific changes in federal and institutional policy could stimulate significant productivity by these people.

- The key ingredient in the management of these scientists and other university scientists is vigorous, creative leadership.

EPSCoR sees universities and their science and engineering departments as resources that are valuable to their regions' agricultural, industrial, and natural economies.[1]

The NSF reports that the "historical concentration of academic R&D funds among the top research universities has remained relatively steady over the past 20 years." The foundation continues:

- In terms of total R&D funding, the share of all academic R&D expenditures received by the top 100 academic institutions decreased from 83% to 80% between 1986 and 1993 and has remained at that level through 2006.
- Only 5 of the top 20 institutions in 1986 were not in the top 20 in 2006.[2]

The federal funding system has been responding too slowly to shifts in the distribution of scientific talent. This debate about the concentration of federal science funding never seems to be resolved. In June 2009, the American Association of Universities, which consists largely of leading research institutions, issued a statement urging that in economically stressed times, federal research funding be targeted to leading universities.[3] Otherwise, they argued, the nation's research standing might suffer.

University Research and the Surrounding Community

The students and citizens of a city or region have a stake in how research is funded and conducted at the local university. For one thing, as discussed in chapter 8, a productive research and teaching program can provide a skilled labor pool and can attract high-tech firms to the area. Also, the university has a responsibility to ensure that campus research is conducted safely.

When the history of science in the 20th century is written, an unlikely figure, Alfred Vellucci of Cambridge, Massachusetts, may figure prominently. The efforts of Cambridge citizens and politicians, spearheaded by Vellucci, to review and assure reasonable precautions in recombinant DNA research being done at Harvard started a domino effect that led to new National Institutes of Health (NIH) standards. Furthermore, the recombinant DNA debate has become a model for how such science policy decisions can be made in the future.

As in many university towns, city politics in Cambridge, Massachusetts, reflect a town-gown split. During the 1960s the Cambridge city council was divided almost evenly between representatives of Harvard and the Massachusetts Institute of Technology on the one hand and representatives of the city's blue-collar working people on the other. More frequently than not, the key vote belonged to Alfred Vellucci. He repeatedly gave outlandish speeches ridiculing Harvard, endearing him to many of the townspeople of Cambridge. However, rumor had it that on the key votes he quietly supported the universities. His rhetorical and political skills combined to make him arguably the most powerful man in Cambridge at that time. In his speeches he proposed, at one time or another, that

- Harvard be requested to move from Cambridge to Waltham, Massachusetts;
- Widener Library (one of the world's great libraries) be converted into a public restroom; and
- the famous Harvard Yard be paved and converted into a public parking lot.

By the 1970s, Alfred Vellucci had become mayor of Cambridge. During this time, biological scientists were poised to launch new investigations into recombinant DNA, research that offered the possibility of incredible intellectual and human payoffs. The Harvard biology department, led by Nobel laureate James Watson, was at the cutting edge of this research. Expressing concern that the DNA research would create "monsters that would crawl over the wall from Harvard and attack the good citizens of Cambridge," Vellucci called hearings to examine the potential threats posed by the research. Nobel laureates and other leading scientists dutifully appeared to discuss the potential value of their research. Vellucci appointed a citizen court to review the evidence.

Does the average citizen have a right to contribute to policy development about science and technology? This question raises the larger question of the role of science and scientists in a democratic society.

Science and the scientific method have transformed our world in wondrous ways, especially since the Renaissance. But some scientific advances also pose threats. Albert Einstein contributed brilliant insights about physics that yielded the horror of atomic weapons as a direct application. The Industrial Revolution transformed economies throughout the world but also brought unparalleled levels of pollution. Insecticides facilitate the production of delicious fruit, but the same spray that protects the orchards causes birth defects among children born to women who live nearby. Some theories about the origins of the AIDS epidemic have attributed it either to scientific research gone awry (recombinant DNA research) or to failed technologies (faulty World Health Organization inoculation programs for other illnesses).

The Founding Fathers could not have anticipated the scientific revolution in the United States. However, Thomas Jefferson, while secretary of the treasury, recognized the importance of science and took time from his other duties to test new inventions. Indeed, while there is a long history of federal concern about science, the government did not enter the science business in a big way until after World War II.

A Brief History of Federal Support for Scientific Research

Federal expenditures for scientific research have been commonplace since the spectacular technical success of the Manhattan Project, which developed the first atomic bombs. In 1945, the case for continued government support of basic scientific research was made by Vannevar Bush, then director of the Office of Scientific Research and Development, and by others. The major science organization to grow out of this federal concern was the National Science Foundation. Through the NSF and various government departments and agencies, federal monies were channeled into universities for research and development activities. In the late 1950s, with the voyage of Sputnik, science education became a national priority, as discussed in the introduction. That period spawned a wide array of measures in support of science education, the National Defense Education Act being a prime example.

By the 1960s, universities were highly affluent, and their scientific activities were generously underwritten by the government. Expansion and growth were widely held objectives for universities and for science in general. The tremendous acceleration of the growth of science following World War II forced a reconsideration of the role of the scientist. Scientists and experts in other disciplines became ubiquitous in the federal government, holding elected and appointed positions alike. Experts also have enormous influence over the research and development industry that focuses on policy analysis. When the newspaper reports the findings of a government study released by the Department of Transportation, or Housing and Urban Development, or Education, few citizens realize that the study was almost certainly done externally, under contract, by a private group of experts. There are many nonprofit corporations and profit-making firms that exist entirely on the basis of such "soft money," otherwise known as grants and contracts, from the federal government. Some of the better-known examples are the Rand Corporation, the Stanford Research Institute, and Abt Associates.

The Ecology of Scientific Research

Both the surrounding community and the surrounding society, as represented by the federal government, are part of the ecology of scientific research. So are the undergraduate and graduate students who are apprentices, learning the scientific method.

America's educational systems present a paradox. While our elementary and secondary schools are weak compared with those of other nations, we have the world's best higher education system. A 2006 article in the *Economist* noted: "Scientists in America will win more Nobel prizes than those in any other country and produce more high-quality academic articles. . . . European intellectual stars will continue to forsake the common rooms of Oxbridge and the cafes of Paris for the research facilities of American academia."[4]

The nature of research changed dramatically in the 20th century. Now, teams of researchers, frequently using expensive equipment, work on massive projects, often under federal funding. Scientific research has become a complex system incorporating psychodynamic, interpersonal, institutional, and economic factors. Consequently, the effective use of management techniques assumes great importance in determining scientific success. The management techniques used to support and evaluate science-building programs may be the critical factor determining which funded enterprises and research projects are productive and which are not. By manipulating those factors and utilizing appropriate management techniques, we can nurture quality research.

A full study of the ecology of scientific research and a full model of the research process must include a number of factors beyond the creativity of the original idea, including the following:

- The background of the researcher: demographic characteristics, academic training, prestige in the department or organization, mentors, pre-PhD productivity, postdoctoral experience, first job, and early career productivity.
- The researcher's entrepreneurial skills: proposal writing, communication with federal officials, opportunities to learn career development from mentors, understanding of the peer review system, ability to articulate ideas, ability to plan research, and ability to evaluate and modify a research program in the light of subsequent success and failures.

- Institutional factors: barriers and stimulants to research such as instrumentation, research assistants, release time, travel funds, and seed money; a critical mass of stimulating colleagues; institutional track record in producing quality research.
- Federal funding policies: emphasis on specific disciplines or subdisciplines, strengths and weaknesses of the peer review process, and possible geographical or institutional biases.

Many policy studies about the research process employ an oversimplified mechanistic model as though the following process occurs: (1) a brilliant scientist has an idea, (2) he or she gets funding and tries the idea, (3) the idea either works or does not work, and (4) technological innovation and industrial productivity result if it does. The new research reality is much more complex, demanding a knowledge and application of management techniques. "Middle-range" sociological theories, such as reference group theory, effectively explicate and enhance our understanding of the research process. For example, I conducted analyses to explore the manner in which "relative deprivation"—which can be likened to being either a big fish in a small pond or a small fish in a large pond—might affect the self-esteem, risk-taking, and productivity of scientists, particularly young ones.[5] Some researchers are more successful in environments where they are the big fish. Others perform better at larger institutions where they are surrounded by a critical mass of excellent peers. New lines of research are needed to explore the many ways in which scientists are inspired and go on to create new knowledge. Kuhn's landmark 1962 book, *The Structure of Scientific Revolutions*, suggests that science progresses through incremental elaborations of an existing "paradigm," or theory, until someone radically redefines the fundamental paradigm of the discipline. Research at the fringes of science may be where new paradigms are incubated.

Proposals and Grant Funding

It is difficult to judge the performance of scientific funding agencies, for, like physicians, they often bury their mistakes. Rejected proposals usually mean doomed projects. If the projects survive rejection and succeed, it is rare that they achieve recognition soon enough to alert the funding agencies that mistakes are being made. In 1978 I was given the Alan T. Waterman Award of the National Science Foundation and the Texas Instruments Foundation Founders' Prize for research that initially had been rejected for funding by the National Science Foundation (NSF), the Department of

Energy (DOE), the National Aeronautics and Space Administration (NASA), and the Department of Defense.

Richard Muller, "Innovation and Scientific Funding"[6]

It became clear as I studied university research that the weak link in the chain of activities from project planning through funding, research, and publication is the point at which a faculty member prepares a grant proposal. Some policy studies have compared the number of awards to the number of proposals received. But the causal chain extends to the proposal submission process itself. It is important both to examine what kinds of scientists are motivated to prepare proposals and to establish the degree to which institutional affiliation facilitates or inhibits this process. How much more likely is the same faculty member to write a proposal at Yale than at a remote state university? If potentially excellent researchers are inhibited from project planning and proposal development because of their institutional affiliation, this has important implications for federal and university policy.

Prior to the 19th century, discoveries and innovations in the fields of science and technology were mostly the result of individual genius and effort. During certain periods in history, nations sponsored technology for military reasons, or wealthy patrons with a keen interest in science supported researchers, or universities engaged in the transmission of scientific knowledge—but discovery was primarily the result of scientific and technical entrepreneurs. Just as the 19th century saw the development and application of the scientific method of inquiry, the 20th century saw vast changes in the nature and structure of technological work. Methodical inquiry led to systematic research; systematic research required team efforts by highly trained specialists from different disciplines; and the housing of these specialists gave rise to the development of university, industrial, and government research laboratories. Innovation thus became specialized, professionalized, and institutionalized. Sponsorship of innovation has fallen mainly to governmental agencies or large corporations who can afford to staff facilities and support personnel, and who have a vested interest in the products and processes developed.

The process of systematic research in both basic and applied areas has positive and negative effects. At the positive end of the scale, the resources needed to conduct investigations, explore alternatives, and develop potential are available. At the negative end, the institutionalization of the process and the methods of obtaining sponsorship can stifle creativity, originality, or inventiveness. This is especially true for emerging nontraditional fields whose advocates and practitioners may not be affiliated with prestigious institutions, or lack the prerequisite list of educational and professional credentials, or are not attuned to the skills involved in soliciting and receiving funding for research and development activities.

In the course of my field research, it was inevitable that some particularly glaring examples of biased or incompetent peer reviews emerged:

- In at least two institutions, scientists said they had submitted proposals that were rejected because their work was found to be infeasible, yet in each case the researcher previously had accomplished the "infeasible" task—making a certain chemical compound. One reviewer had commented: "This work has been attempted at many large centers. What makes —— think he could do it at —— University?" But the scientist had already been doing this work.
- One scientist gave us a review that included the following statements: "I do not know much about the subject matter discussed in this proposal. . . . Because I am not in the field of the investigator, I prefer not to comment." However, the reviewer had rated the proposal "good" and "fair," and these ratings may have been considered in the foundation's decision to reject the proposal.
- One proposal was rejected after a reviewer commented, "This work would require many other scientists to travel to —— state, and everyone knows there is no air travel to ——," a patently absurd statement.
- A reviewer commented about a proposal that was rejected: "If it were really good chemistry, MIT would have done it already."
- And the grand prize winner: "This has never been done before."
- In two different states the same saga was described: scientists who submitted a proposal were turned down, moved to more prestigious institutions, resubmitted the same proposal, and received funding.

In the first draft of the classic 1945 report *Science: The Endless Frontier*, principal author Vannevar Bush proposed that university research funding by the federal government be based on the British system, where funds were awarded proportional to student enrollment. By the final draft, he had been persuaded to insert the phrase "peer review," a change that had profound implications for elitism in university research funding and for the productivity of university research funding.[7]

The Corporate Context

The Bayh-Dole act of 1980 revolutionized the relationships between university and industry, solidifying corporations' role as key players in the ecology of scientific research. The act made it possible for both professors and their universities to receive income from patents developed from the research of professors. The popular sports drink Gatorade is a great example of how universities can profit from innovation from their own campuses; because of its success on the commercial market, the University of Florida receives "$11 million per year for the faculty alone from Gatorade royalties."[8] In 2004, American colleges and universities earned over $1 billion in revenue from intellectual property licensing. Out of more than 3,000 institutions nationwide, just two (Columbia and NYU) accounted for 20 percent of that revenue. More than half of the total revenue went to eight institutions.[9]

The participation of academics in medical research has remained strong over the past decade. However, industry funds an increasing proportion of this influential medical research, especially randomized controlled trials, most of which are now funded exclusively by industry.[10] Increasingly, American scientists who collaborate with industry sign nondisclosure agreements. The purpose of these agreements is to protect the companies from having new findings and discoveries stolen and used by their competitors. However, the agreements can also undermine the integrity of scientific research. Boyce Rensberger, director of the Knight Science Journalism Fellowship program at MIT, says, "I used to argue that we could count on academic scientists to tell us the truth because they were independent and honest, but nowadays I'm not so sure."[11]

The Science Development Program

Through research conducted at the National Board on Graduate Education (NBGE), housed within the National Research Council, my colleagues and I evaluated the major NSF funding program of the 1960s and 1970s, Science Development, which often was referred to as the "Centers of Excellence" program. Science Development grew out of the perennial debate about the distribution of federal research funds, a debate that had first surfaced in the 1940s when the NSF was created.

The purpose of the Science Development program was to increase the scientific productivity and excellence of institutions outside the top 20. The project began with the assumption that distribution of research funds should maximize scientific productivity and quality. Some argued strenuously that it was obvious that research funds should go to the best scientists, most of whom were to be found in the top institutions. Others argued for geographical and institutional equity, but with awards still granted on the basis of excellence. They contended that funding decisions betrayed an elitist bias towards the top universities (roughly half the federal funding in each scientific discipline was being awarded to the top 20 institutions) and that scientific talent was more widely dispersed. Furthermore, they argued that creative undergraduate and graduate students attending lower-ranked universities were less likely to enter scientific research careers unless they worked with professors who were active, funded researchers.

Science Development embodied three innovative concepts: (1) a strong emphasis on geographical dispersion of the funded institutions; (2) funding of "second-tier" institutions (those not yet considered excellent), together with deliberate exclusion of those universities in the top 20; and (3) funding via institutional rather than project support. Science Development involved three subprograms:

1. University Science Development (USD), under which 31 institutions received $177 million. The central philosophy of emphasis on systemic institutional support, not project support, was embodied in the USD subprogram.
2. Special Science Development, under which 11 universities that had been denied USD support received limited funding (about a million dollars each) for a few promising departments.
3. The Departmental Science Development subprogram, which awarded grants (about $600,000 each) to individual departments that showed promise but were housed in lower-ranked universities, often urban universities.

There often is a discontinuity between Washington politics and evaluation rigor. The top 20 research universities were excluded from funding under this program. But NSF administrators were quite cautious and never actually named the excluded universities. Leading research universities searched for measures that would define them as outside the top 20 so that they could then compete for these (then) massive funds. The most successful of those leading institutions was the University of Washington, which, by my calculations, then ranked about eighth in the nation, and was funded under USD. The largest Departmental Science Development grant went to Yale! And so it goes in Washington.

My research team at the National Research Council was tasked with conducting a comprehensive evaluation of Science Development, including field-visit case studies of funded and nonfunded institutions and statistical analyses of quantitative data. For the field visits, we assembled teams of scientists. Recruiting scientists to assist the National Academy of Sciences is not difficult; for example, our team included national leaders like John Bardeen, who twice won the Nobel Prize for physics and was co-inventor of the transistor; policy experts; institutional leaders; and evaluation specialists. We interviewed university administrators, faculty and staff, and, sometimes, students at each university. We amassed a database of longitudinal indicators about graduate departments in physics, chemistry, and mathematics (the disciplines receiving funding) and a control field, history. We collected information from primary sources and from existing databases on a number of indicators of science department structure, functioning, and success. The data spanned a 15-year period including the years before, during, and after Science Development funding.

We built multivariate models to assess the unique impact of Science Development funding on important outcomes while controlling for baseline differences between institutions. The project report appeared as a book, *Science Development: An Evaluation Study*, along with an accompanying National Board on Graduate Education policy report.[12] While we were objective and included criticisms (e.g., there was virtually no concern expressed for undergraduate education and the scientific pipeline at the funded institutions), we reported dramatic changes and quantum leaps in research quality at a number of institutions. For example, the program funded three universities in close proximity in North Carolina, catalyzing the development of the Research Triangle in that state.

A 1971 article about the NSF summarized the impact of the Science Development program by noting that its institutional support and fellowship grants elevated "graduate schools and departments to top rank. Privately, NSF was given much of the credit for the development of the highly regarded astronomy department at the University of Arizona, the mathematics department of Louisiana State University, and the physics departments at Rutgers and the University of Oregon, among others. Arizona alone received $7.2 million from NSF over the past five years."[13]

At the University of Arizona, Science Development funds were used to bolster several physical science departments, notably astronomy. A major portion of the award to Arizona was used for the construction of a 90-inch reflector telescope on Kitt Peak, near Tucson. The Arizona astronomy department was not considered for ranking in a national evaluation in the mid-1960s; however, by the time the next report was published, in 1970, it was ranked fifth in the country. Subsequently, physicists at the University of Arizona created quite a stir in the scientific community when they published findings that called into question portions of Einstein's theory of relativity. Whether this scientific success should be directly attributed to the Science Development program is debatable, but it certainly provides a dramatic example of the potential for science building at institutions that previously had mediocre or weak research programs.

There were many positive effects of the program: brilliant new faculty hired, graduate enrollment increased, buildings constructed, a telescope built, an exciting interdisciplinary institute established, and so forth. In some institutions the progress was dramatic, particularly given the short time span involved, and some might argue that the progress of these schools alone would justify the program, particularly when compared with the results of most federally funded activities.

When we compared extremely successful grants with less successful ones, certain characteristics stood out, namely the strength of the central administration: *grants tended to be most successful at universities that had a strong and dynamic leader before, during, and after the grant.* This was particularly true of USD grants. Continuity in office was another important factor. Upheavals were sometimes observed when an institutional leader who had been instrumental in acquiring the grant left prematurely (from the point of view of Science Development). A strong leader committed to the university was vital to the grant's success. Even though all concerned tried to use multiple objective criteria in awarding grants, it is clear in retrospect that one cannot underestimate the influence of personality. Whether a grant proposal originated with a department or with the central administration, one person usually played a driving role. For example, at one public institution the chairman of the physics department prodded other department heads to write their sections of the proposal. Both in the preparation of proposal and in the success (and occasional failures) of the grant, the presence of a strong individual was central. The person who receives project support from the federal government may become more committed to his research and less dependent on or committed to the institution, while the person who successfully administered a Science Development grant had a stronger vision of what the institution or department might become. The chief lesson to be learned from this comparison of successes and failures was that future funding programs should support the person who has demonstrated *commitment to the institution,* not a person who will leave or who will favor a pet area.

Another component of success, and one that is associated with strong central leadership, is an overall development plan for the university. For many of the successful schools, the creation of a Science Development proposal amounted to carving out a section of an existing plan prepared as the result of extensive self-study; the Science Development funds contributed to an overall balanced effort. In addition, these schools tended to be the ones that were best managed overall and relatively strong financially as a result. Successful institutions with these characteristics were able to use program funds to help strengthen non-funded departments and improve undergraduate education.

The Experimental Program to Stimulate Competitive Research (EPSCoR)

The NSF launched the Experimental Program to Stimulate Competitive Research (EPSCoR) in the early 1980s to rectify what some perceived to be a continuing inequitable distribution of NSF monies.[14] States, rather than regions, were chosen as the basis for the experiment, largely because the regional governmental structure of the country is less developed than state structures. States were also deemed more capable of supporting and carrying on the development projects after the cessation of NSF support; regions have no comparable mechanisms for doing so. NSF's Experimental Program is designed to facilitate self-improvement in selected states. The program's goal is to improve the ability of scientists in participating states to compete successfully for NSF and other competitive federal research programs by

- improving the quality of research being conducted in participant states,
- increasing the number of nationally competitive scientists and thereby the federal funding in participating states, and
- effecting long-term gains in the research environment in participant states.

EPSCoR places responsibility for program development and execution with state ad hoc committees selected in part by the NSF. To initiate the program, the NSF selected the states eligible for participation and assembled the initial members of the ad hoc committees. The program was divided into two phases: phase A, planning awards; and phase B, implementation of awards. The process of selecting the original participant states was based largely on prior federal funding activity to these states. The NSF wanted to build the scientific research capabilities of states whose scientists had fared very poorly in previous grant and contract competitions. Seven states were deemed eligible for participation: Arkansas, Maine, Montana, North Dakota, South Carolina, South Dakota, and West Virginia.

During the planning phase the state committees were responsible for assessing the quality of science within their state; identifying the state's scientific problems, resources, and options for improvement; and developing an implementation plan to improve the quality, strength, and competitive status of science within the state. These nine-month planning grants resulted in full implementation proposals with which the seven states competed for funding under phase B. Each state could participate in the program for five years and receive up to $3 million over that period, depending on the plan it developed. NSF support would decrease gradually from approximately 90 percent of the total in the first year to 40 percent in the fifth year, with the states expected to provide matching funds to make up the difference. As envisioned and encouraged by the NSF, the seven implementation proposals resulting from the planning grants represented a concentrated effort to increase the ability of scientists in those states to compete successfully for federal research funds. The program descriptions reflected each state's response in light of its scientific, socioeconomic, geopolitical, and academic environments. They also represented various philosophical viewpoints on how best to achieve that objective.

Monitoring and evaluation plans were to be included in each state's proposal. Those assessment activities would begin when the implementation award was made and would continue for the duration of the program. The ad hoc state committee played a key role in local oversight of the program's functioning. After two years of implementation, the state committee was to conduct a review of the program to enable the state and the NSF to evaluate progress and to provide an opportunity to modify, and perhaps even end, the plan if progress was not satisfactory. Upon agreement of the participating scientists, the state committee, and the NSF, the program would be continued for the remainder of the five years.

I directed a policy analysis review of EPSCoR, with a focus on the early planning and implementation activities in these seven states. The results were reported in *Strengthening Academic Science*.[15] The progress made by the five initial recipient states—Arkansas, Maine, Montana, South Carolina, and West Virginia—was substantial, and Congress subsequently authorized EPSCoR funding in additional states and territories. Moreover, the success of EPSCoR spawned parallel programs in other federal agencies, including the Department of Energy. Furthermore, the European Economic Community has conducted a planning study to see whether and how the EPSCoR model could be applied in the less scientifically developed countries of Europe.

EPSCoR Today

The first EPSCoR grants provided each of the five funded states a budget of $3 million per year over a five-year period. EPSCoR clearly filled a national need, and the program has grown over the ensuing 30 years. In 2008, the NSF EPSCoR budget was $115 million. Moreover, Congress expanded the EPSCoR concept to five other federal agencies. Table 9.1 shows the 2008 funding for these agencies. Federal funding for all EPSCoR programs in 2008 totaled $402.3 million.

EPSCoR today has three funding components:

1. A program to improve the research infrastructure. This bears the closest resemblance to the original program. An important component which funds the development of the cyber infrastructure has been added.
2. Funding for deserving research not otherwise eligible. Sometimes proposals to various divisions of NSF are judged to be meritorious through peer review but cannot be funded because of limited resources in the appropriate division. Through co-funding with the relevant division, EPSCoR can ensure that this important scientific work goes forward.
3. Support for workshops, conferences, and outreach to the EPSCoR states.

TABLE 9.1
EPSCoR funding for federal agencies, 2008

Federal agency[a]	Funding (millions of dollars)
National Institutes of Health	$220.5
National Science Foundation	115.0
Department of Agriculture	19.2
Department of Defense	17.1
NASA	15.5
Department of Energy	15.0
Total	$402.3

Source: EPSCoR/IDeA Foundation, Budget, www.epscorfoundation.org/about/budget.

[a] The Environmental Protection Agency also has an EPSCoR program, but it received no funding in 2008.

Recently announced EPSCoR awards funded projects in six states, including, for example, research on the following topics:

- Current threats to the biodiversity and ecosystem of the Hawaiian islands, including invasive species and human activities. Principal investigator James R. Gaines observes: "The development of high performance computing models and new 3-D visualization systems gives us information that can help us make informed decisions regarding public policy and land use that will benefit both the people and environment of Hawaii."
- A South Carolina study of biofabrication, using computer technology and biological innovations to engineer functional tissues and organs. According to principal investigator Jerry Odom, "This is a very collaborative and cooperative project whose end goal is to grow organs. Just think how important that could be to people waiting for transplants."[16]

Elsewhere, EPSCoR-funded researchers are working on a variety of other projects. For instance:

- Researchers in South Dakota are trying to develop less costly solar cells made of plastic, using carbon-based semiconductors, as opposed to inorganic silicon-based semiconductors, which are expensive.
- In Mississippi scientists have developed a method to use penicillin to coat surgical tools and medical implants, an innovation with the potential, pending clinical studies and commercial development, for saving 90,000 lives per year in America alone by preventing surgical infections.
- Researchers in Vermont are developing innovative sensors and transmitters that can detect when stress is becoming intolerable for a variety of objects, such as huge mining trucks and human knees. According to Vermont engineer Steve Arms, "By calculating the amount of fatigue that a component has been exposed to, it is possible to repair or replace the component only when required."[17]

Each EPSCoR state initiative involves a collaborative effort among colleges and universities across the state and usually also involves business, industry, and stand-alone laboratories. As in the first days of the program, each state's effort is guided by a board comprising university, business, and government leaders. The program now includes 25 eligible states and two territories. However, even today, while EPSCoR states contain 25 percent of the nation's research or doctoral institutions, they receive less than 15 percent of total federal research and development funding.[18]

In Conclusion

Encouraging and improving university research is an essential part of expanding America's capabilities in STEM fields. We must optimize both the quality of the research and the opportunity for all students, regardless of geographical location, to learn from experienced researchers. As the National Academy of Sciences report *Rising above the Gathering Storm* recommended, we must renew and support America's commitment to the research necessary for economic, national security, and social progress.[19] Only through increased investment in the STEM fields—and in the students and scholars within them—can the United States reclaim its competitive advantage in the high-tech global economy.

Conclusion

> Full many a gem of purest ray serene
> The dark unfathom'd cave of ocean bear:
> Full many a flower is born to flush unseen,
> And waste its sweetness on the desert air.
> Thomas Gray, "Elegy Written in a Country Churchyard"

Ultimately, this is a book about power in U.S. society. Better educational opportunities—especially in the STEM fields—give students more power over their own lives.

The inequality in our economy is staggering. Throughout history, people with power have controlled the means for the weak and disenfranchised to improve their positions. In the emerging global economy, upward social mobility will require technical skills and knowledge. Our young people can acquire those skills only through STEM education.

Women, poor people, and disadvantaged minority students are consistently discouraged from studying science and mathematics, the very subjects that would give them access to power, influence, and wealth. Sometimes this discouragement takes the form of overt racism or sexism. More frequently, good intentions (combined with pernicious expectations) suggest that disenfranchised groups cannot master mathematics and science. The cycle continues when students themselves incorporate this false expectation, lower their own self-assessments, and limit their aspirations.

The mentoring insights, evaluation techniques, teaching strategies, and other reforms outlined in this book represent substantial potential improvement in STEM education. But the most critical change involves raising expectations. Teachers must realize that virtually every student—regardless of gender, ethnicity, or economic status—can master mathematics and science. Parents must realize this. Most importantly, students themselves must understand this.

Examples I have given underscore how negative expectations about a group's mental capabilities can limit that group's access to quality education. Other examples show how some individuals' extraordinary achievements have forced the elite group to reconsider these prejudices. The towering mathematical achievements of Ramanujan silenced those who believed that East Indians possessed limited intellectual capabilities. Karl Pearson's wrong-headed views on the intelligence of Jews were eventually refuted. Still, theses about limitations imposed by innate intelligence were taken seriously at one time. The damage done when such ideas are embraced cannot easily be undone.

Aptitude is a limited concept, at best. Moreover, aptitude has been just as destructive through its misuse as it has been helpful through its use. Measurement of a person's aptitude (for example, by an intelligence test, which gives us only an approximate measure of aptitude) tells us very little about what that person will learn or achieve in the workforce. Aptitude is not a pure, inherited set of traits that exist in a vacuum; rather, intelligence and ability occur in a context. For this reason, modern psychometricians, led by Howard Gardner, have developed the notion of "multiple intelligences," such as linguistic, musical, spatial, and interpersonal intelligence.

The circumstances in which aptitude is the overriding factor are rare. The unquestioned genius of Mozart, or Ramanujan, or Einstein transcends normal boundaries. At the other extreme, there are profoundly handicapped children and adults whose ability to learn is severely limited. But for the great majority of the human population, intelligence is only one small factor affecting what they learn and achieve. Hard work is also a factor. Effective teaching is, too. So is structuring effective environments for learning.

It is time we stop excluding students from educational opportunities because of outmoded ideas about aptitude and who can learn. A student should not be denied the opportunity to study higher mathematics and science solely because his or her parents are poor. Nor should teachers, parents, or other persons of authority write off the learning potential of young women or minority students. We must end the cycle of negative expectations and wasted talent in this country.

We should expect virtually all students to be able to learn and master science and mathematics. In fact, we should expect students to excel in these subjects. In order to succeed, students may have to work harder, but the benefits will be incalculable. We can no longer afford to write off these students on the assumption that only exceptional individuals can do mathematics and science. This assumption is empirically unjustified, unfair to the students, and a danger to the future of our economy.

In attempts to diagnose why so many avoid or are discouraged from pursuing technical fields, mathematics was increasingly revealed to be the fulcrum or filter. Many scientists, mathematicians, and teachers believe that only certain people can learn mathematics. As a result, by junior high school, many students regard mathematics as a boring, irrelevant, and difficult subject. It does not help matters that some teachers who are skilled at mathematics are not skilled at communicating with people.

In our society some people practically brag about their inability to do mathematics. This attitude is lethal for educational and career development. Too many college students choose their majors and careers on the basis of a careful avoidance of mathematics courses, limiting their career options and reinforcing stereotypes about aptitude or intelligence. Even students whose future occupation will minimally involve technology must make important decisions in their personal lives that involve mathematics, science, and technology, such as decisions made in the voting booth.

"More Like Us"

The Chinese philosophy of Taoism says that leaders must look to themselves, not to others, to lead effectively. The same can be said for education reform. I believe there is much we can learn from the educational systems of other countries. But we should not simply try to imitate them, nor should we lose sight of the unique strengths of our system.

Consider John Wooden's coaching strategy. To date he has been the most successful college basketball coach in history. His UCLA teams won seven consecutive NCAA titles and won ten national championships in 12 years. Wooden focused totally on leading his team to play to their potential. He spent little time studying the other team.[1]

In the early 1990s, James Fallows studied the educational systems of several Asian nations where students earned higher achievement scores than American students. He concluded that simply imitating the educational systems of these other nations would not improve U.S. scores. Rather, we should build on our unique strengths to become "more like us."[2]

Talent in Unexpected Places

Not every sixth grader who will make future creative contributions to scientific research is found in a special track or program for talented children. Not all highly successful electrical engineers took AP calculus in high school. Similarly, the greatest scientific discoveries of the next 25 years may not be made by tenured faculty members at our leading universities, or by professional employees at the major technology companies.

- A breakthrough medical advance may come from a medical doctor working alone in his clinical practice.
- The latest economic treatise may be written by a man sitting alone at a table in a museum.
- New scientific theories may be constructed by a clerk in a government office.

I grew up in the middle of the 20th century. The intellectual, political, economic, and scientific world I grew up in had been shaped largely by three innovative thinkers in precisely the positions I have just described.

- Sigmund Freud's theories totally revolutionized not only psychiatry and the treatment of psychopathology, but also how we thought about the functioning of all adults. He introduced such concepts as the unconscious and sexual repression and totally changed how people thought about what it means to be human. Freud was a practicing psychiatrist, a medical doctor in Vienna, who developed his theories based on what he learned from his patients through a process he developed that he called "psychoanalysis," utilizing especially a technique he developed called "free association."
- Karl Marx published his ideas about the nature of capitalism and about how economic rewards should be distributed in *The Communist Manifesto.* He developed those ideas and wrote his book while sitting alone at a table in the British Museum in London.
- Albert Einstein's insights about energy, time, and motion brought about a paradigm shift in physics and in our understanding of the universe. He developed these ideas while working full time as a clerk in a patent office in Switzerland. He took the job in the patent office after being turned down for several university positions.

None of these three giants had a federal grant. None of them had a research team. Each worked alone.

Win-Win Education

Barriers and restrictions to STEM education will continue to damage our economy and our culture. In the information age, we lose nothing and gain everything by providing opportunities to virtually every young person.

A rising tide lifts all boats.

In mathematics game theory, there is the zero-sum game, in which you win and I lose, and the nonzero-sum game, in which it is possible for both of us to win. In a classic work on international negotiation, *The Strategies of Conflict*, Richard Schelling argues that diplomacy is a nonzero-sum game. Everybody wins. I believe that STEM education is also a nonzero-sum game. The more we can expand this subject knowledge to students, the better off everyone in our society will be. The global economy, too, will benefit. There is no longer any justification for rationing education to the privileged few.

Early in my career I was a computer programmer and systems analyst. We programmed mainframe computers (which filled a large room) in a language called FORTRAN (short for "formula translation") and in assembly and machine languages. Both the computer program and the data were stored on punch cards. We saved our work on magnetic tapes, which were a key form of storage. Contrast the computers of the past with the desktop or laptop computer you use today. (Or, for that matter, with your cell phone!) The speed and memory available on current electronic devices is enormous compared with the largest and fastest computer available at Harvard University in the 1960s.

The development of educational opportunities for American students throughout history has been similar. Only 35 years ago the career opportunities for people studying STEM disciplines were limited, just as there was limited magnetic core computer memory. Today, however, career opportunities in the STEM fields are vast, just like the memory on our personal computers.

Talent Can Come from Anywhere

In his book *The Black Swan*, Nassim Nicholas Taleb argues that the major events that challenge our worldview are not predicted by conventional forecasting methods and statistical techniques.[3] Just as everyone believed that all swans were white until they discovered a black swan, change or new discoveries force us to reconsider all our assumptions.

Taleb illustrates his argument with powerful metaphors, such as the Thanksgiving turkey. Using its three years of life as evidence, the turkey, perhaps applying multiple regression analysis or time series analysis, might expect that humans would always feed and treat him well. That turkey is in for a rude awakening come Thanksgiving!

To further describe the futility of conventional forecasting methods when unexpected events occur (and Taleb believes all major events are unexpected), he asks us to consider two scenarios. In scenario A, 100 people are lined up in the middle of a football stadium and weighed. The addition of one more person, no matter how heavy that person is, will not change the mean weight significantly. But in scenario B, the same 100 people are lined up in the stadium and report their net worth. Adding Bill Gates to the lineup will dramatically change the mean net worth.

Taleb argues that conventional statistical techniques based on Gaussian normal distributions are useless when they encounter a black swan situation. He offers an alternative: the fractal mathematics developed by Benoit Mandelbrot.

Black swan reasoning suggests that we should educate as many young people as possible in the STEM disciplines, provide wide access to college, and distribute federal research support widely. You simply cannot predict—especially using outmoded, useless aptitude concepts—who will be creative and innovative.

Second Chances

Americans believe in second chances. In many countries, including countries whose students outperform American students on international assessments, poor test scores in secondary school, elementary school, and perhaps even preschool may limit a student's future prospects. In the United States it is possible to recover from early failure. In fact, our national history has many stories of triumph following failure. President Harry Truman preferred that all potential cabinet appointees had suffered some major failure in their past.

The American notion of second chances is consistent with what we know from both research and experience about how human lives unfold. Consider Jim Clark. As an adolescent, Jim did not adjust well to high school. He flunked out of several high schools and was kicked out of others. After dropping out, he joined the U.S. Navy. There he was given an intelligence test. Jim Clark scored so poorly that the navy decided that the only task he was suited for was mopping the decks. During his navy tenure, Jim became friends with someone who saw in him talent and encouraged him to pursue his education. Eventually Jim earned a bachelor's degree, a master's degree, and a PhD. Jim Clark moved to Silicon Valley and started not one, but two highly successful companies: Silicon Graphics and Netscape. One of the forces behind the information revolution that has transformed the world economy, he is now one of the world's wealthiest men.

All this from a man who failed repeatedly in high school and who was judged by the navy to be capable only of swabbing the deck.

Sociologist, priest, and novelist Andrew Greeley reflects about the meaning of second chances in his own life: "What are the illuminations in my story? It is risky to reduce poetry to prose, fiction to nonfiction. But I would like to think that the illumination in my story is that we live in a cosmos that is finally, however oddly, implacably forgiving; that it is never too late to begin again; that there are always second (and more) chances; that it is possible, Ulysses-like, to go home again; that we will all be young again and all laugh again; that love is always and necessarily renewable; and that life is stronger than death."[4]

The Case for Traditional Education

Why learn mathematics, science, and history when increasingly the world of work is driven by computer technology? After all, some of the fastest growing occupational categories involve fixing computers and running complex software systems. Why bother with a traditional undergraduate education? Wouldn't it make more sense to train in a discipline with relevance to the job market?

A young person in Las Vegas might ask the same questions and assume that a college education isn't needed if there are lucrative jobs in hospitality. But will the same kinds of jobs be available 10 or 20 years from now? If not, what resources will this person have to shift gears and move into another line of work? Computing technology changes rapidly. Tomorrow's technology jobs may not exist today.

Ray Kurzweil has studied the rate of change of computer technology, and he believes that its growth pattern is exponential. (This is merely a mathematician's way of describing a growth rate that is really fast and accelerating.) In *The Age of Spiritual Machines*, Kurzweil suggests changes and dilemmas our society might confront within the next quarter century, if not sooner.[5] He speculates about what might happen when computers not only out-think humans, but develop consciousness.

Verbal interaction with computers may soon be commonplace. That is, I could verbally state my analysis request to the computer. The computer—really, the software—could verbally ask me questions about anything that wasn't clear, and so forth. In such a scenario, many of today's computer-based jobs would vanish.

In the 1960s, when I was a computer specialist, the computing center seemed like a temple of knowledge and I felt like a priest or rabbi. Now most of that information and knowledge is available to any student at the click of a mouse. The same transformation could happen to many of today's computer specialist jobs.

A century ago, long-distance communication occurred by telegraph. Telegraph operators worked in special locations. They were experts in communication through a special "language," Morse code. Telephones changed that. Up until the 1950s, long-distance calls were handled by an operator. Touch-tone dialing made the telephone accessible to all, even very young users. Now three-year-olds can place a telephone call on a cell phone merely by pushing a preprogrammed button. Times change. Technologies change. The prevalence and nature of technology-based occupations change. A traditional education in the liberal arts prepares people to understand the limits of technology, to place technology in context.

The "appropriate technology" (AT) movement has underscored the limits to the benefits of technological change. As is the case with many social or scientific movements, differing streams of influence were integrated and galvanized by one charismatic visionary. E. F. Schumacher was a German-born, Oxford-trained economist who left academia to gain experience in business, farming, and journalism. In 1965 he established the Intermediate Technology Group in England. In 1977 he published *Small Is Beautiful: Economics As If People Mattered*,[6] which, according to one reviewer, was "a work of stunning brilliance that seems certain to become a modern classic. It is one of those books that can wipe clean the windows through which we look at the world so that we can see many things that formerly were either vague or invisible."[7] In this marvelous book, Schumacher argues:

> If we ask where the tempestuous developments of world industry during the last quarter-century have taken us, the answer is somewhat discouraging. Everywhere the problems seem to be growing faster than the solutions. This seems to apply to the rich countries just as much as to the poor. There is nothing in the experience of the last twenty-five years to suggest that modern technology, as we know it, can help us to alleviate world poverty, not to mention the problem of unemployment, which already reaches levels like 30 percent in many so-called developing countries, and now threatens to become endemic also in many of the rich countries. In any case, the apparent yet illusory successes of the last twenty-five years cannot be repeated. . . . So we had better face the question of technology—what does it do and what should it do? Can we develop a technology which really helps us to solve our problems—a technology with a human face? (139–40)

Schumacher labeled his work "Buddhist economics" and said that "the very start of Buddhist economic planning would be a planning for full employment, and the primary purpose of this would in fact be employment for everyone who needs an 'outside job': it would not be . . . the maximization of production" (53).

To illustrate the idea that small is beautiful, Schumacher argues, for example, that the maximum desirable size for a city is about half a million people. Beyond that, nothing is added but problems and human degradation. A counter example has been cited by AT proponents: In the 19th century the Russians built the world's largest bell, weighing 250 tons. It broke the first time it was rung. Even though it was useless, it was still proudly displayed in the Kremlin. After all, it was the world's largest bell.

While appropriate technology usually is small technology, AT proponents emphasize technology *appropriate* to the environment, which does not always mean small. Appropriate technology conserves energy and resources. Sometimes appropriate technologists simply urge society to be more efficient. For example, Schumacher commented in an interview with *The Futurist* magazine on the "immensity" of transport in our society: "If I travel from London to Glasgow on one of the big motorways, I find myself surrounded by huge lorries carrying biscuits from London to Glasgow and I look across the other lane and I find an equal number of lorries carrying biscuits from Glasgow to London. Any impartial observer from another planet would come to the conclusion that biscuits, in order to achieve proper quality, must be transported at least 500 miles."[8]

Higher Expectations

One of the best solutions for strengthening STEM education is to raise our expectations for students. All of us—teachers, parents, students, everyone—must recognize every child's intellectual potential for mastering science and mathematics. Consider this statement by David Peters:

> [My grandfather] had only a third grade education, but learned veterinary medicine by correspondence. He knew the importance of education and he had an analytic mind. When he was about 90, I asked him how he would calculate the amount of corn that was in a circular crib. His response is what I call the "Herman Peters' Theorem on Pi." He drew a circle on a piece of paper. "Suppose the crib was 20 ft. across," he said. "Put a square around it and divide into fours." "That makes 400 square feet. If you cut each square in half, the resulting diamond is exactly half the area, or 200 square feet. Since

> the circle is about half way between the two squares, I would say that it must be about 300 square feet." For those of you who are geometrically challenged, the exact answer is 100 pi or 314 square feet. He had computed pi to within 4.5% from first principles. What would he have done, I thought, with a college education?[9]

Throughout this book I have identified ways to educate more students in the STEM fields: college calculus workshops, junior high school science courses with practical applications, incentives for qualified teachers, and the efforts of the National Science Foundation. None of these initiatives is a magic bullet solution to our mathematics and science education quagmire. First we must raise expectations that all students, regardless of race, gender, or social standing, are capable of learning math and science.

Daryl Smith, a colleague at Claremont Graduate University, notes that Major League Baseball invests heavily in players from the Dominican Republic. Young Dominicans with athletic potential are recruited as early as elementary school into special camps and workshops. They are trained, guided, and encouraged. Recognizing this country's pool of talent, Major League Baseball has signed many of these players to its teams' rosters. Players from the Dominican Republic represent the largest percentage of foreign-born players, with nearly 100 baseball players active in the league today.[10] Do Dominicans have a genetic advantage when it comes to playing baseball, or is their high level of skill and success a function of the intensive training they receive?

Talent is distributed more or less the same in every sociological group. Talent is not the principal determinant of who succeeds in mathematics and science, or of who succeeds in life.

Reform Should Focus on Process

Too often, attempts to reform STEM education seek a simple, concrete solution. Change the curriculum. Hire teachers who excelled in college. Select a charismatic school principal. Recruit the most talented students. While each of these strategies should be part of the solution, taken together, they fall far short of implementing deep, permanent improvement in STEM education. True reform requires transforming the way we teach, learn, and lead.

Almost 20 years ago, Sheila Tobias observed: "Materials development remains the darling of the science education community. Why the nearly exclusive focus on instructional materials? Is it because these are products that give educational reformers and their paymasters something to show? *Or is it that science education reformers, with some notable exceptions, don't know what else to do?* . . . The temptation is to solve a problem with a product (or . . . a project) because, for innovators and funders alike, the messy, intensely local alternatives are harder to conceptualize."[11]

I would argue that we now know what else to do. And it's time to do it.

Changing the Way We Teach and the Way We Lead

> Never again will you be a group. (Odds against, trillions to one.) We've been together thirty hours, here in this room whose gaseous cylinders emend the erratic window light. Those spritzes of autumn the neo-Venetian neo-Gothic windows admit . . .
>
> Though I will see some of you again, will write many of you letters of recommendation—for years to come—may even, God knows, teach your children (if you have them soon), may, some day, in Tulsa, or West Hartford, see you when your present beauty is long gone, I know that what counts for us is over . . .
>
> Teaching.
>
> Richard Stern, "Wissler Remembers"[12]

In 1989 in hearings before the Subcommittee on Postsecondary Education of the U.S. House of Representatives, I was asked to testify about mathematics and science education. Throughout the day a number of other scientists, policy analysts, and experts testified, too. However, the most memorable statement that day was made not by a policy wonk or an education expert, but by Kent Kavanaugh, a high school teacher from Missouri. Earlier that year Kavanaugh had received a national award from the president of the United States in recognition of his outstanding work as a science teacher. This is part of what he said to the committee: "The month before I left for Washington, DC, to receive the Presidential Award, I averaged forty hours a week at my job—not my teaching job, my second job as an analytical chemist for the Mobay Corporation of Kansas City. I must work at two or more jobs to make ends meet and to try to pay for college tuition for my son who graduates this spring. Because of this, I often come to school physically and mentally worn out. I am not the exception, I am the rule."[13]

Even the best teachers need a supportive work environment. Furthermore, we can educate, train, and develop excellent teachers.

Quality leadership is rarely defined by one brilliant, charismatic individual. The answer of successful leadership is not in *who leaders are*, but rather in *what they do.* The same is true for teachers. Yes, charismatic leaders can transform organizations and nations. And, yes, charismatic teachers often transform the lives of their students. But most teachers and leaders do not possess such rare talents. Nonetheless, they can become gifted leaders and teachers by learning how to be effective. James Stigler and James Hiebert ask:

> What does it mean to focus on teaching? It means, first of all, becoming aware of the cultural routines that govern classroom life, questioning the assumptions that underlie these routines, and working to improve the routines over time. It means recognizing that the details of what teachers do—the particular questions teachers ask, the kind of task they assign students, the explanations they provide—are the things that matter for students' learning.
>
> But, more than that, focusing on teaching means recognizing that all of these details of teaching are *choices* teachers make.[14]

Being an effective teacher or leader involves understanding the ecology, norms, and culture of the organization. Jean Lipman-Blumen, a leading scholar who studies leadership, argues that leaders must "engage the synergistic effects of such leadership: seeing and making connections where others don't; viewing diversity as a valued reservoir of resources; harnessing the ego to the high purposes and burdens of the group; translating passion for individuals into compassion for the group; and setting oneself and others on a lifelong search for authentic experiences and greater understanding."[15] She notes that leaders' repertoire of such skills allows them "to draw upon a broader spectrum of behaviors than they may be accustomed to, most importantly a set of political or 'instrumental' styles that use the self and others as instruments for accomplishing goals. Because connective leaders use these instrumental styles ethically and altruistically, they can integrate the otherwise centrifugal forces of diversity and inter-dependence."[16]

The challenge is not to find talented leaders and managers, although that will help. The challenge is not to find talented teachers, although that will help. The challenge is to create a learning environment—an ecology—that fosters student achievement. The issue isn't so much who is selected to be a teacher, principal, or superintendent. The issue is what these educators do. Research and experience have shown us the strategies that work with students. We know that teachers and leaders can learn these strategies and implement them. *Professional development is crucial.* Even experienced teachers can learn to improve and transform their teaching.

The Joy of Science

Scientific work is not only socially relevant; it is also a labor of love for many researchers. In an article about politics and the academy, Elie Kedourie cites Yeats's poem "An Irish Airman Foresees his Death":

> Nor law, nor duty bade me fight,
> Nor public men, nor cheering crowds,
> A lonely impulse of delight
> Drove to the tumult in the clouds.

"Through the power of the poet's genius," says Kedourie, "these lines illumine, by analogy, the life of scholarship, and the particular fulfillment which is its reward."[17]

Laura Andersson, an Oregon biochemist, expressed similar sentiments:

> In retrospect, I cannot honestly say that I became a scientist solely because of my teachers. I had marvelous and dedicated teachers—and others who were less so. However, any motive force came primarily from within—an inescapable compulsion to ask questions and explore their answers. . . . The societal pressures against scientific careers for women still exist—and were even more prevalent when I was an undergraduate. But they never mattered, or were never a factor sufficient to deflect me from the sheer joy and unequalled exhilaration that are the rewards of working in the discipline.[18]

We can create an environment in which our citizens are active participants in the high-tech economy of the 21st century. We have the knowledge, and the power, to transform American education through

- committed and responsible *leadership*;
- *evaluation* of all proposed educational initiatives and balanced assessment of student achievement;
- recruiting, rewarding, and respecting excellent *teachers*—and providing them with ongoing professional development and renewal;
- high *expectations* that all students will succeed;
- closing the achievement gap;
- recognizing the importance of *mentors*;
- increasing and facilitating *access to college*; and
- *strengthening university research* at the undergraduate and graduate levels.

Appendix

International Assessments of STEM Achievement in the Twentieth Century

ASSESSMENTS IN THE 1960S AND 1970S

Has the achievement of American students declined during the past 40 years, or have they always scored poorly? Findings from the original surveys by the International Association for the Evaluation of Educational Achievement (IEA), for which planning began in 1966, are revealing. A 12-country feasibility study had been conducted in 1962, and a study of mathematical achievement in 12 countries was published in 1967.[1] The results of the mathematics achievement testing done in the 1965 international assessment are summarized in table A.1. Out of 70 questions U.S. math students got an average of 13.8 questions correct, considerably below the nearest score among the other countries in the study (Australia, whose students averaged 21.6 correct answers).

Three areas—science, literature, and reading comprehension—were studied between 1970 and 1972 (with reports published in 1973); three more areas—English, French as a foreign language, and civic education—were studied in 1974 and 1975. Nineteen countries participated in the science study, although data from Chile, India, Iran, and Thailand were analyzed separately from data from the developed countries. In 1976, David Walker published a volume summarizing the results from all six subject surveys, which were conducted in 21 countries. One year later, Richard Wolf published a report about the IEA results as they related specifically to the United States.[2]

The results from the science tests showed American students faring worse as they grew older. Among 10-year-olds, the youngest students, the United States did reasonably well. The mean score of 17.7 placed it 4th out of the 12 developed countries from which scores were obtained. Among 14-year-olds, the United States ranked 7th out of the 14 developed countries that reported data. However, by the senior year of high school, the United States was in *last place* in science. In fact, the

TABLE A.1
Results of the 1965 international mathematics assessment

Country	Mean score[a]	
	Math students	Non-math students
Australia	21.6	
Belgium	34.6	24.2
England	35.2	21.4
France	33.4	26.2
Germany	28.8	27.7
Israel	36.4	
Japan	31.4	25.3
The Netherlands	31.9	24.7
Scotland	25.5	20.7
Sweden	27.3	12.6
United States	13.8	8.3

Source: T. Husen, ed., *International Study of Achievement in Mathematics* (New York: John Wiley, 1967), p. 24.

[a]Score is given as an average number of correct responses on a 70-item test.

U.S. score was not substantially higher than that achieved by Thailand, one of the developing countries. The United States also ranked last in the grand total score, which included some extra advanced questions.[3]

Even 40 years ago, American high school seniors were performing dismally on international science achievement tests. Today's low scores probably do not represent a substantial decline in the quality of American schools.

What happens if we compare the best U.S. students in these assessments from the early 1970s with the best students from other countries? Mean scores for each country were produced for the top 1 and 5 percent of high school seniors. Looking at the top 1 percent of each country's students, the United States no longer places last, but rather 9th out of 14. Looking at the top 5 percent, the United States places 8th. In short, when we control the bias introduced by the differing percentages of students studying science in various countries, American high school seniors move from last place to the middle of the pack.

Field testing in the six subject surveys conducted in the early 1970s involved 258,000 students, 50,000 teachers, and 9,700 schools. Why, one might ask, did the researchers embark on such a venture? "We conceived of the world as one big educational laboratory where a great variety of practices in terms of school structure and curricula were tried out. We simply wanted to take advantage of the international variability with regard both to the outcomes of the educational systems and the factors which caused differences in those outcomes."[4] This pioneering IEA report on science education in 19 countries took place

> at a time when the nature of science education and its contribution to a general education, as distinct from a specific training, was coming under close scrutiny in many parts of the world. Traditional patterns, affecting both subject content and learning methods, were giving way to new programs, often under the stimulus of curriculum projects organized on a vast scale and employing new curriculum reform techniques. It is fair to say that the study was made at a critical stage in the history of science teaching and the results obtained may influence considerably the direction of future progress.[5]

Survey data collected on elementary and secondary science education in the 1970s show that the proportion of secondary school students studying science in the United States was relatively low compared with other counties, including Chile, Finland, Hungary, and Thailand, among others. For example, 43 percent of students in the United States took science in high school, whereas 100 percent of the students in Thailand did so. In fact, the only country that registered a lower percentage was England, whose educational system forced students to make major curriculum choices earlier in their school studies than did other countries.[6]

Richard Wolf observed that "students in the United States performed in close accord with the amount of opportunity to learn what was being tested. *The implication here is that increased opportunity to learn specific material will lead to higher test performance.*"[7]

The evidence appears to indicate that when the proportion of the school-age population still attending school is controlled, U.S. students perform at or just below the middle of the pack. Furthermore, while it is difficult to estimate accurately the impact of test item bias—and the data on bias vary from study to study—it is possible that bias might increase the true U.S. ranking somewhat. However, even correcting for these biases in an effort to help the U.S. ranking only improves its standing from disastrous to mediocre.

ASSESSMENTS IN THE 1980S

Most Americans first became aware of international assessments of educational achievement in the 1980s. The IEA conducted an international assessment in mathematics during the 1981–82 school year. Five topics were included: arithmetic, algebra, geometry, statistics, and measurement. Eighth-grade students in Japan scored ahead of all other countries in all five subject areas. U.S. students scored a little above average in calculation, a little below average in problem solving, about average in algebra, and far below average in geometry.[8]

Twelfth-grade students were assessed on six topics: number systems, sets and relations, algebra, geometry, elementary functions and calculus, and probability and statistics. For this age group, Hong Kong students scored best in each topic, with Japan a close second. U.S. calculus students scored about average, which is not surprising considering that fewer students take calculus in the United States than in other countries, but American precalculus students fell far below average. The United States ranked in the bottom quarter in some areas of mathematics and was the lowest-ranked overall among advanced industrialized countries.[9]

The authors of this IEA report had some recommendations for improving education in the United States: "With respect to form, the excessive repetition of topics from year to year should be eliminated. A more focused organization of the subject matter, with a more intense treatment of topics, should be considered."[10] They also recommended less repetition of arithmetic and greater inclusion of geometry, probability and statistics, and algebra in the junior high school curriculum.

To control for differing proportions of young people still in school and of young people still taking mathematics as high school seniors in different countries, a special analysis was made of the achievement scores in algebra of the top 1 and 5 percent. The top American students scored lower than the top students from any other country for which data were available.

In the eighth-grade population, there was a close correspondence between the percentage of test items taught in a country's curriculum, which is considered an "opportunity-to-learn" measure, and the achievement test scores. The United States ranked slightly below the middle on opportunity to learn. While this situation was true in algebra, all countries scored relatively low on the opportunity to learn geometry.

The authors of this international assessment considered five explanations often offered for the low achievement level of American students—explanations that they labeled "deceptive"—such as class size and time for mathematics instruction. They concluded that the curriculum was the culprit. They reported that their study showed the U.S. mathematics curriculum to be underachieving "in its goals, in its strategies and in its expectations for students." The U.S. eighth grade curriculum, they said, "is much more like a curriculum of the last years of elementary school while that of Japan, and many other countries, resembles that of the first years of secondary school. At the twelfth grade level, the U.S. curriculum is much more like that of early years of secondary school while the curriculum of most other countries is more like that of beginning college level."[11]

In other words, our expectations for U.S. students are too low.

Perhaps the most widely publicized assessment of the era was the science assessment conducted by the IEA between 1983 and 1986. Its preliminary report, *Science Achievement in Seventeen Countries*, changed many Americans' perception of the U.S. educational system.[12]

This IEA study examined education in Australia, English-speaking Canada, England, Finland, Hong Kong, Hungary, Italy, Japan, South Korea, the Netherlands, Norway, the Philippines, Poland, Singapore, Sweden, Thailand, and the United States. Separate assessments were conducted of students at three levels: 10-year-olds (typically grade 4 or 5), 14-year-olds (typically grade 8 or 9), and students in the final year of secondary school (typically grade 12). While the 10- and 14-year-olds took a general test of science competence, the students tested during their final year of secondary school were assessed separately in biology, chemistry, and physics. All told, 220,848 students from 9,808 schools participated in the assessment.

As the age level increased, the U.S. students moved from the middle of the pack to close to the bottom or dead last, depending on the subject. With respect to the 14-year-olds in England, Hong Kong, Italy, Singapore, and the United States, "the lowest scoring children were scoring at chance level, indicating that from the test's point of view, they were *scientifically illiterate*."[13] American students in grade 9 had about the same level of achievement as students in grade 7 in Japan and Korea.[14] At all levels, male students scored higher than females (more about this in chapter 2), and this gender difference was greater for 14-year-olds than for 10-year-olds.

Samples of some of the questions asked give a flavor for this assessment. A sample item from the test given to 10-year-olds reads as follows:

> The sun is the only body in our solar system that gives off large amounts of light and heat. Why can we see the moon?
> a. It is reflecting light from the sun.
> b. It is without an atmosphere.
> c. It is a star.
> d. It is the biggest object in the solar system.
> e. It is nearer the Earth than the Sun.

Only 66 percent of American students chose the correct answer (a), ranking American students 5th out of 15. Countries with higher percentages of students answering correctly were Finland (68%), Hungary (68%), and Sweden (70%). Poland had the lowest score on this item, with only 41 percent of students choosing the correct answer.

When the same question was asked of 14-year-olds, the number of U.S. students answering correctly increased to 70 percent. Despite the improvement, the relative position of the United States slipped to 12th, as more countries than before exceeded that score. Australia (slightly higher), Canada (73%), Hong Kong (77%), the Netherlands (74%), the Philippines (72%), Poland (72%), and Singapore (89%) all scored better than the United States.

The next question was also administered to both groups:

> Paint applied to an iron surface prevents the iron from rusting. Which one of the following provides the best reason?
> a. It prevents nitrogen from coming in contact with the iron.
> b. It reacts chemically with the iron.
> c. It prevents carbon dioxide from coming in contact with the iron.
> d. It makes the surface of the iron smoother.
> e. It prevents oxygen and moisture from coming in contact with the iron.

The correct answer, e, was selected by 46 percent of American 10-year-olds. Countries' rates of correct responses varied from a high of 70 percent for Finland to a low of 32 percent for Singapore. The United States placed above only four other countries: England, the Philippines, Hong Kong, and Singapore. When this question was asked of 14-year-olds, 66 percent of American students answered correctly. However, American students surpassed only those from the Philippines, where 51 percent gave the correct answer. Fourteen-year-olds from Hungary performed best on this question, with 91 percent giving the correct answer.

It has been argued that many apparently superior educational systems in other countries (for example, in Japan) require a considerable amount of memorization, while the American educational system places a high priority on creativity, cognitive flexibility, and the ability to interpret complex information. Consider a question administered to students in the final year of secondary school, requiring interpretation, not merely feeding back memorized information:

> In order to obtain two crops in one growing season, a farmer planted some seeds which he had harvested the previous week but the seeds failed to germinate. What can be concluded from this observation?

a. The farmer did not provide the right conditions for germination.
b. The seeds needed a longer period of maturation.
c. The farmer had not removed inhibiting substances.
d. The seeds required a period of low temperature.
e. The data are inadequate for a conclusion to be reached.

The correct answer (e) was selected by 53 percent of students from the United States. Singapore scored considerably higher (66%). The only other country whose students scored higher than the American students was England, with 55 percent, only marginally higher than the American score. Particularly low scores were earned by the students from Finland (22%), Hungary (23%), Italy (27%), and Norway (28%). In other words, in this assessment, American students performed well on an item that required analytical reasoning and critical thinking.

In some countries, including the United States, there is high variability among schools, while in other countries the quality of schools is relatively uniform. To account for this discrepancy, the authors of the study computed the percentage of schools in each country that scored below the lowest school in the highest-scoring country, which was Japan. In the United States, 38 percent of the schools attended by 10-year-olds had mean scores that fell below that of the worst school in Japan. Among 14-year-olds, the top-scoring country was Hungary. Thirty percent of U.S. schools had mean scores below that of the worst school in Hungary. For students in the last year of secondary school, three countries tied for first place in chemistry, and 48 percent—nearly half—of U.S. schools scored below the worst schools in those countries. Hong Kong was the top country in physics, and *89 percent* of U.S. schools did more poorly than the worst school in Hong Kong. Finally, *98 percent* of U.S. schools scored below the worst school in Singapore, which was the top country in biology.

Some scholars have criticized the IEA study's methodology. First, the samples may not be truly comparable because the United States retains a higher percentage of students in school in the senior year than do other countries. In addition, the tests inadvertently favored small countries with highly centralized curricula.

One critic concluded: "As a science teacher, I must ask myself the following question after reading the IEA study: Can a single twenty-four- to thirty-item test possibly be an accurate index to the curriculum in my classroom, in all other American science classrooms, and in the science classrooms of sixteen other countries? The degree of disagreement in my answer should be proportional to the degree of skepticism that I exercise as I interpret IEA's results."[15]

I present data that counter these criticisms below. The consistently poor performance of American students on different tests over a 40-year history cannot be ignored.

IEA researchers provided detailed data about the sampling at each age level. Examination of the data shows that virtually 100 percent of the children aged 10 and 14 were in school in each country. In the senior year of high school, 90 percent of U.S. students were in school, while the percentages for the other countries were considerably lower—71 percent for Canada, 20 percent for England and Hong Kong, 18 percent for Hungary, 63 percent for Japan, and 17 percent for Singapore.

The percentages of U.S. students who were studying biology (6%), chemistry (1%), and physics (1%) in the senior year (the group that was actually tested) are considerably lower than the percentages in most other countries. Conversely, 66 percent of U.S. students did not study science in their senior year, higher than any other country that reported data.

With respect to the second criticism, the IEA researchers emphasized that these were international tests: "The development of all instruments was a collaborative effort involving all educational systems in the study at that time. This is essential for the validity of the cross-national comparisons that are made. A common curriculum grid was developed using the curricula of all systems participating in the study: Items were provided by all systems for the measurement of particular cells in the grid; any new item was trialed in at least five different systems."[16]

In 1989 a dozen countries and Canadian provinces participated in an international assessment of achievement in mathematics carried out by the Educational Testing Service. Table A.2 shows the rankings in this assessment of 13-year-olds; U.S. students came in last.

TABLE A.2

Ranking of countries and provinces participating in the 1989 international assessment of mathematics achievement: 13-year-olds

1. Korea	7. New Brunswick, French
2. Quebec, French	8. Spain
3. British Columbia,	9. United Kingdom
4. Quebec, English	10. Ireland
5. New Brunswick, English	11. Ontario, French
6. Ontario, English	12. United States

Source: *World of Differences: An International Assessment of Mathematics and Science* (Washington, DC: Educational Testing Service, 1989).

ASSESSMENTS IN THE 1990S

A study in the 1990s measured the mathematics and science achievement of 9-year-olds in 14 countries and of 13-year-olds in 20 countries. Here, I focus on the results for 13-year-olds, who took a 76-item mathematics test and a 72-item science test.

In the introduction to their report, the authors underscored the importance of these international comparisons in the emerging high-tech global economy:

> Some might say that a study that compares the United States with Slovenia or England with São Paulo Brazil, is inappropriate or irrelevant. Indeed, education is, in fact, imbedded in each society and culture, and performance should not be studied or described without considering the important differences from country to country. The life of a thirteen-year-old in a rural Chinese community is very different from that of his or her peer growing up in a middle-class Paris apartment. And yet, these two young citizens may well meet in the global marketplace twenty years from now. And if they do, chances are they will rely on the mathematics and science they learned in this decade to succeed in the complex business and technological environment of 2012.[17]

The findings from the United States were not encouraging: 13-year-olds scored below those from every other country except Ireland and Jordan. Among the countries scoring higher than the United States were Spain and Slovenia.

In mathematics, U.S. students performed below those from all the other assessed countries except Jordan. When I conducted research in Arkansas, I often heard the phrase, "Thank God for Mississippi." While Arkansas ranked dismally in many quality of life measures, sometimes 49th out of 50 states, it could rest assured that Mississippi would rank lower. Perhaps we Americans should be saying, "Thank God for Jordan."

Upon reviewing the results from these international assessments, President George H. W. Bush and the nation's governors set a goal: that the United States would become number one in the world in mathematics and science by the year 2000. We did not achieve that goal.

Results published by the National Assessment of Educational Progress (NAEP) in 1994 showed some slight improvement in science proficiency compared to an earlier NAEP study, which may have been due to reform efforts. Yet, after reviewing the full range of results from this study, Mark Musick, president of the Southern

Regional Education Board, commented: "While the trend is up in science and math—which is heartening—'up' doesn't seem to be very high. . . . Virtually all thirteen- and seventeen-year-olds can read, write, add, subtract and count their change. But as one moves up the scale toward slightly more complicated tasks, student success falls off rapidly."

Notes

INTRODUCTION

Epigraph: Quoted by Gerald W. Bracey, "America, the 97-Lb. Weakling?" *Principal Leadership (High School Ed.)* 6, no. 3 (2005).

1. William J. Jorden, "Soviet Fires Earth Satellite into Space; It Is Circling the Globe at 18,000 M.P.H.; Sphere Tracked in 4 Crossings over U.S.," *New York Times*, Oct. 5, 1957, p. 1.

2. A Harvard University research team evaluated the impact of New Math in the Brookline, Massachusetts, school system. As a novice graduate student, I played a small role in this evaluation project in 1960, observing classrooms where the reforms were implemented.

3. Palminteri Chazz, *A Bronx Tale*, directed by Robert de Niro (New York, 1993).

4. John Allen Paulos, *Innumeracy: Mathematical Illiteracy and Its Consequences* (New York: Hill & Wang, 1988).

5. Anne C. Lewis, "Endless Ping-Pong over Math Education," *Phi Delta Kappan* 86 (2005).

6. Sheila Tobias, *Overcoming Math Anxiety* (New York: Norton, 1993).

7. Christopher Cerf and Victor Navasky, *The Experts Speak: The Definitive Compendium of Authoritative Misinformation* (New York: Villard, 1998), p. 191.

8. Ibid., p. 170.

9. Linda Darling-Hammond, "Standards, Accountability, and School Reform," *Teachers College Record* 106 (2004): 1047–85.

10. *Rising above the Gathering Storm: Energizing and Employing America for a Brighter Economic Future* (Washington, DC, National Academy of Sciences, 2007), p. 17.

11. John Eichinger, conversation with author, 1989.

12. Lewis, "Endless Ping-Pong."

13. Motoko Akiba, Gerald K. LeTendre, and Jay P. Scribner, "Teacher Quality, Opportunity Gap, and National Achievement in 46 Countries," *Educational Researcher* 36 (2007): 369–87.

14. Linda Darling-Hammond, "Teaching as a Profession: Lessons in Teacher Preparation and Professional Development," *Phi Delta Kappan* 87 (2005): 237–40.

15. Ibid., p. 239.

16. Carolyn M. Callahan et al., "TIMMS and High-Ability Students: Message of Doom or Opportunity for Reflection?" *Phi Delta Kappan* 81 (2000): 790.

17. John-Paul Sartre, preface to Frantz Fanon, *Wretched of the Earth* (Paris: F. Maspero, 1961).

18. Uri Treisman, "A Study of the Mathematics Performance of Black Students at the University of California, Berkeley" (PhD diss., University of California–Berkeley, 1985).

19. Martin V. Bonsangue and David E. Drew, "Long-Term Effectiveness of the Calculus Workshop Model" (report to the National Science Foundation, April 1992); Richard J. Light, "The Harvard Assessment Seminars, Second Report, 1992: Explorations with Students and Faculty about Teaching, Learning, and Student Life" (Cambridge: Harvard University Graduate School of Education and Kennedy School of Government, 1992).

20. Robert B. Reich, *The Work of Nations: Preparing Ourselves for Twenty-First-Century Capitalism* (New York: Alfred A. Knopf, 1991).

21. James S. Coleman et al., *Equality of Educational Opportunity* (Washington, DC: U.S. Government Printing Office, 1966).

22. Robert Rosenthal and Lenore Jacobson, *Pygmalion in the Classroom* (New York: Holt, Rinehart, & Winston, 1968).

23. David E. Drew, *Strengthening Academic Science* (New York: Praeger, 1985).

24. Ibid.

25. Quoted in Parker J. Palmer, *The Courage to Teach: Exploring the Inner Landscape of a Teacher's Life*, 10th anniversary ed. (San Francisco: John Wiley and Sons, 2007), p. 46.

CHAPTER 1: AMERICA'S PLACE IN THE WORLD

Epigraph: National Commission on Excellence in Education, *A Nation at Risk: the Imperative for Educational Reform* (Washington, DC: U.S. Department of Education, 1983).

1. *Rising above the Gathering Storm: Energizing and Employing America for a Brighter Economic Future* (Washington, DC, National Academy of Sciences, 2007), p. 17.

2. Thomas L. Friedman, *The World Is Flat: A Brief History of the Twenty-First Century* (New York: Farrar, Straus and Giroux, 2005), p. 237.

3. Ibid., p. 275.

4. Lester Thurow, *Head to Head: The Coming Economic Battle among Japan, Europe, and America* (New York: William Morrow, 1992), p. 309.

5. Jack Ewing et al., "The Rise of Central Europe," *Business Week*, Dec. 12, 2005, p. 52.

6. T. R. Reid, *The United States of Europe: The New Superpower and the End of American Supremacy* (New York: Penguin, 2004), p. 1.

7. Claire Berlinski, *Menace in Europe: Why the Continent's Crisis Is America's Too* (New York: Crown Forum, 2006), p. 311.

8. *Rising above the Gathering Storm*, p. 14.

9. James H. Johnson Jr. and John D. Kasarda, "People and Jobs on the Move: Implications for U.S. Higher Education," in *Redefining Student Success: The Challenges and Implications of Extending Access* (New York: The College Board, 2006), pp. 47–54.

10. David Halberstam, *The Reckoning* (New York: William Morrow, 1986), pp. 270–71. Hereafter cited by page number in the text.

11. Paul DeHart Hurd, *New Directions in Teaching Secondary School Science* (Chicago: Rand McNally, 1969), p. 1.

12. National Science Board, *Science Indicators 2006* (Washington, DC: National Science Federation, 2006).

13. David Goodstein, "The Science Literacy Gap: A Karplus Lecture," *Journal of Science Education and Technology* 1 (1992).

14. Sheldon Glashow, "Burning Questions: Losing the Future," *ABC News Special* (1988).

15. George F. Will, "Building a Wall against Talent," *Washington Post*, June 26, 2008, p. A19.

16. *Rising above the Gathering Storm*, p. 10.

17. Margaret Spellings, "Building America's Competitiveness: Examining What Is Needed to Compete in a Global Economy," testimony before the House Committee on Education and the Workforce, 109th Cong., 2nd sess., 2006.

18. R. E. Yager and J. E. Penick, "Perceptions of Four Age Groups towards Science Classes, Teachers, and the Value of Science," *Science Education* 70 (1986): 360.

19. Mariann Lemke and Patrick Gonzales, *U.S. Student and Adult Performance on International Assessments of Educational Achievement: Findings from the Condition of Education 2006*, NCES 2006-073 (Washington, DC: U.S. Department of Education, National Center for Education Statistics, June 2006).

20. Robyn Baker and Alister Jones, "How Can International Studies Such as the International Mathematics and Science Study and the Programme for International Student Assessment Be Used to Inform Practice, Policy and Future Research in Science Education in New Zealand?" *International Journal of Science Education* 27, no. 2 (2005): 147.

21. Lemke and Gonzales, *U.S. Student and Adult Performance on International Assessments*.

22. Ibid, p. 15.

23. Ibid, p. 24.

CHAPTER 2: THE ACHIEVEMENT GAP

Epigraph: Stephen Jay Gould, *The Mismeasure of Man* (New York: Norton, 1981), p. 60.

1. Richard Tapia, commencement speech, Claremont Graduate University, May 2008.

2. Richard Tapia, "Broadening Participation: Hiring and Developing Minority Faculty at Research Universities," *Communications of the ACM* 53, no. 3 (2010): 33–34.

3. Deborah Taylor and Maureen Lorimer, "Helping Boys Succeed: Which Research-Based Strategies Curb Negative Trends Now Facing Boys?" *Educational Leadership*, Dec. 2002. pp. 68–70.

4. National Science Foundation, Division of Science Resources Statistics, "Undergraduate Enrollment," in *Women, Minorities, and Persons with Disabilities in Science and Engineering: 2002*, NSF 03-312 (Arlington, VA: 2003), at www.nsf.gov/statistics/nsf03312/c2/c2s5.htm#c2s5l1.

5. Jacqueline E. King, "Gender Equity in Higher Education: 2006," *On Campus with Women* 36, no. 3, at www.aacu.org/OCWW/volume35_3/feature.cfm?section=2.

6. *Gender Differences at Critical Transitions in the Careers of Science, Engineering, and Mathematics Faculty* (Washington, DC: National Science Foundation, 2009), pp. 153, 154, 158. Also reported by K. Harmon, "Why Aren't More Women Tenured Science Professors?" *Scientific American*, July 19, 2009.

7. Ibid., p. 144.

8. Helen S. Astin, "Citation Classics: Women's and Men's Perceptions of Their Contributions to Science," in *The Outer Circle: Women in the Scientific Community*, ed. H. Zuckerman, J. R. Cole, and J. T. Bruer (New York: Norton, 1991), p. 68.

9. Ibid.

10. Ben A. Barres, "Does Gender Matter?" *Nature* 442 (July 2006): 133–36.

11. Maria Klawe, Telle Whitney, and Caroline Simard, "Women in Computing—Take 2," *Communications of the ACM* 52, no. 2 (2009): 69.

12. Committee on Maximizing the Potential of Women in Academic Science and Engineering, National Academy of Sciences, National Academy of Engineering, and Institute of Medicine, *Beyond Bias and Barriers: Fulfilling the Potential of Women in Academic Science and Engineering* (Washington, DC: National Academies Press, 2007), pp. 2–4.

13. Clifford Adelman, *Lessons of a Generation: Education and Work in the Lives of the High School Class of 1972* (New York: Jossey-Bass, 1994).

14. Ibid., p. 63.

15. K. D. Rappaport, *Rediscovering Women Mathematicians* (Washington, DC: Educational Resource Information Center, 1978).

16. Marti H. Rice and William M. Stallings, "Florence Nightingale, Statistician: Implications for Teachers of Educational Research" (paper presented at the annual meeting of the American Educational Research Association, San Francisco, Apr. 16–20, 1986).

17. Asha Gopinathan, "Spotlight on Invisible Women," *Science*, Jan. 28, 2005, p. 522.

18. S. G. Brush, "Women in Science and Engineering," *American Scientist* 79 (Sept.–Oct. 1991): 406.

19. Sheila Tobias, *Overcoming Math Anxiety*, rev. ed. (New York: Norton, 1993), p. 74.

20. Ibid., p. 78.

21. Cheryl Ooten and Kathy Moore, *Managing the Mean Math Blues: Math Study Skills for Success* (Upper Saddle River, NJ: Pearson Education, 2010), p. 130.

22. Brush, "Women in Science and Engineering," p. 415.

23. E. Falconer, quoted ibid., p. 24.

24. Carla C. Johnson, "An Examination of Effective Practice: Moving toward Elimination of Achievement Gaps in Science," *Journal of Science Teacher Education*, May 19, 2009, p. 289.

25. "Football—African American athletes—Blacks in Sports: 1947–1992: The Legacy," *Ebony*, August 1992.

26. Douglas Hartmann, "Rush Limbaugh, Donovan McNabb, and 'a Little Social Concern,'" *Journal of Sport and Social Issues* 31, no. 1 (2007): 45–60.

27. Steve Chapman, "Black QBs, QED: The End of an NFL Myth," *National Review Online*, Oct. 3, 2003, www.nationalreview.com.

28. Roy O. Freedle, "Correcting the SAT's Ethnic and Social-Class Bias: A Method of Reestimating SAT Scores," *Harvard Educational Review* 73, no. 1 (2003): 1–43.

29. Ibid., p. 16.

30. Ibid., p. 5.

31. Roy O. Freedle, "How and Why Standardized Tests Systematically Underestimate African Americans' True Verbal Ability and What to Do about It: Towards the Promotion of Two New Theories with Practical Applications," *St. John's Law Review* 80 (2006): 183, 187.

32. Bryan A. Brown, "The Politics of Public Discourse: Discourse, Identity, and African Americans in Science Education," *Negro Educational Review* 56, nos. 2–3 (2005): 205–20; emphasis added.

33. Daryl G. Smith and Gwen Garrison, "The Impending Loss of Talent: An Exploratory Study Challenging Assumptions about Testing and Merit," *Teachers College Record* 107, no. 4 (2005): 629–53.

34. Ibid., p. 637.

35. Ibid., p. 647.

36. American Association of University Women, "Improving Girls' and Women's Opportunities in Science, Technology, Engineering, and Math," AAUW Public Policy and Government Relations Dept., May 2010.

37. Mark H. McCormack, *What They Don't Teach You at Harvard Business School* (New York: Bantam Books, 1984), pp. xv–xvi.

38. Stephen J. Gould, *The Mismeasure of Man* (New York: Norton, 1981).

39. The possible connection between biology, genes, gender differences, and achievement in mathematics, science, and other areas is examined thoroughly by A. Fausto-Sterling in *Myths of Gender: Biological Theories about Women and Men* (New York: Basic Books, 1985).

40. See, e.g., Howard Gardner, *Frames of Mind: The Theory of Multiple Intelligences* (New York: Basic Books, 1983).

41. J. R. Flynn, "Massive IQ Gains in 14 Nations: What IQ Tests Really Measure," *Psychological Bulletin* 101 (1987): 171–91.

42. J. R. Flynn, *What Is Intelligence? Beyond the Flynn Effect* (New York: Cambridge University Press, 2009).

43. Malcolm Gladwell, *Outliers* (New York: Little, Brown, 2008).

44. David W. Galenson, *Old Masters and Young Geniuses: The Two Life Cycles of Artistic Creativity* (Princeton, NJ: Princeton University Press, 2006).

45. Elazar Barken, *The Retreat of Scientific Racism* (New York: Cambridge University Press, 1992), p. 157.

46. Richard J. Herrnstein and Charles Murray, *The Bell Curve: Intelligence and Class Structure in American Life* (New York: Free Press, 1994).

47. Ibid., p. 618. Herrnstein and Murray were working with variables that are accurate to one or two digits, occasionally three, yet they reported beta coefficients and other statistics with eight digits.

48. Robert Kanigel, *The Man Who Knew Infinity: A Life of the Genius Ramanujan* (New York: Charles Scribner's Sons, 1991), p. 3.

49. Ibid.

50. E. H. Neville, quoted ibid., p. 336.

51. Jawaharlal Nehru, quoted ibid., p. 354.

52. G. H. Hardy, "A Mathematician's Apology," in *The World of Mathematics*, ed. J. R. Newman (New York: Simon and Schuster, 1956), pp. 2027–38.

53. R. J. Sternberg, "What Should We Ask about Intelligence?" *American Scholar* 65 (1996): 217.

CHAPTER 3: EFFECTIVE LEADERSHIP, CAREFUL EVALUATION

1. Peter Drucker, "The Age of Social Transformation," *The Atlantic* 274 (Nov. 1994): 53–80, at www.theatlantic.com/ideastour/markets-morals/drucker-full.html.

2. Ibid.

3. Peter Drucker, *The New Realities* (New York: Harper and Row, 1989), pp. 3–4.

4. Ibid., pp. 174–75.

5. Ibid., pp. 3–4.

6. Peter F. Drucker, *Managing the Nonprofit Organization: Principles and Practices* (New York: HarperCollins, 1990), p. 134.

7. M. D. Lemonick, "Drano for the Heart," *Time*, Nov. 17, 2003, pp. 60–61.

8. M. Specter, "A Reporter at Large: The Vaccine," *New Yorker*, Feb. 3, 2003.

9. D. Reeves, *Accountability in Action: A Blueprint for Learning Organizations* (Denver: Center for Performance Assessment, 2000).

10. Gerald N. Tirozzi, "Taking Charge of High School Reform," *Leadership* 34, no. 4 (2005): 8–10.

11. Trish Williams et al., *Similar Students, Different Results: Why Do Some Schools Do Better? A Large-Scale Survey of California Elementary Schools Serving Low-Income Students* (Mountain View, CA: EdSource, 2006), p. 1. Hereafter cited by page number in the text.

12. Linda Darling-Hammond and Olivia Ifill-Lynch, "If They'd Only Do Their Work!" *Educational Leadership* 63, no. 5 (2006): 8–13.

13. Ibid.

14. Karin Chenoweth, *"It's Being Done": Academic Success in Unexpected Schools* (Cambridge: Harvard Education Press, 2008), pp. 11–12. Hereafter cited by page number in the text.

15. Enrique Medina, "Effective High School Reform: Meeting the Academic Needs of Urban High School Students" (PhD diss., Claremont Graduate University, 2009), abstract. Hereafter cited by page number in the text.

16. Michael Pressley et al., "How Does Bennett Woods Elementary School Produce Such High Reading and Writing Achievement?" *Journal of Educational Psychology* 99, no. 2 (2007): 222.

17. Lisa Snell, "The Agony of American Education: How Per-Student Funding Can Revolutionize Public Schools," *Reason*, April 2006, at http://reason.com.

18. Ibid.

19. Lisa Snell, "Meet Arlene Ackerman: The Woman Who Shook Up San Francisco's Schools," *Reason*, April 2006, at http://reason.com.

20. Kristen A. Graham, "Superintendent Arlene Ackerman Named Top Urban School Leader in the U.S.," *Philadelphia Inquirer*, Oct. 22, 2010.

21. Quoted in Joseph B. Platt, *Harvey Mudd College: The First Twenty Years* (McKinleyville, CA: Fithian Press, 1994), pp. 10–11.

22. Ibid., p. 10.

23. Charles A. Wagner, *Four Centuries and Freedoms* (New York: E. P. Dutton, 1950), p. 23.

24. Ibid., p. 11.

25. Ibid., p. 19.

26. This section is based in part on a set of guidelines that Martin Bonsangue and I drew up in July 2009 for congressional aides who were preparing a bill about STEM education.

27. Robert Pirsig, *Zen and the Art of Motorcycle Maintenance* (New York: William Morrow, 1974), p. 184.

28. William Spady, "The Paradigm Trap: Getting Beyond No Child Left Behind Will Mean Changing Our 19th-Century, Closed-System Mind-Set," *Education Week*, Jan. 10, 2007.

29. Geoffrey C. Ward and Dayton Duncan, *The West: An Illustrated History* (New York: Clarkson Potter, 2005), chap. 2.

30. Peter Drucker, personal communication, ca. 1985.

CHAPTER 4: TOP-NOTCH TEACHERS

Epigraph: Peter Drucker, in conversation with Bruce Rosenstein, as quoted at "Drucker Apps" on the Drucker Institute Web site, www.druckerinstitute.com/druckerapps/20090226/index.html.

1. Motoko Akiba, Gerald K. LeTendre, and Jay P. Scribner, "Teacher Quality, Opportunity Gap, and National Achievement in 46 Countries," in *Educational Researcher* 36, no. 7 (2007): 369.

2. Lisa Loop, DeLacy Ganley, and Anita Quintanar, "Examining Teacher Candidates' Experience and Attitudes: Using Baseline Data in Longitudinal Performance Studies" (paper presented at the 8th Annual Hawaii International Conference on Education, Honolulu, Jan. 6–10, 2010).

3. Ibid.

4. I serve as principal investigator or co-principal investigator on National Science Foundation grants to CGU for both the Noyce and Math for America programs.

5. Maria Klawe, personal communication, Oct. 2009.

6. L. Vanderkam, "Making Math Pay: Good Teachers Appear When the Price Is Right," *USA Today*, Education Forum, Jan. 29, 2008.

7. Jim Simons, speech at Harvey Mudd College, Apr. 27, 2010.

8. Parker J. Palmer, *The Courage to Teach: Exploring the Inner Landscape of a Teacher's Life*, 10th anniversary ed. (San Francisco: John Wiley and Sons, 2007), p. 1.

9. D. E. Drew, "Seeing the Forest for the Trees," *Change* 18, no. 4 (1986): 10.

10. John Hattie, *Visible Learning: A Synthesis of Over 800 Meta-Analyses Relating to Achievement* (New York: Routledge, 2009), pp. 238–39.

11. Ibid., pp. 159–60.

12. Ibid., p. 259.

13. Iris Weiss, quoted in J. Raloff, "U.S. Education: Failing in Science?" *Science News*, Mar. 12, 1988, pp. 165–66.

14. Richard J. Murnane and Randall J. Olsen, "The Effects of Salaries and Opportunity Costs on Length of Stay in Teaching: Evidence from North Carolina," *Journal of Human Resources* 25, no. 1 (1989): 106–24.

15. Linda Darling-Hammond, "Preparing Our Teachers for Teaching as a Profession," *Education Digest* 71, no. 4 (2006): 22–27.

16. Linda Darling-Hammond, "Standards, Accountability, and School Reform," *Teachers College Record* 106, no. 6 (2006): 1047–85.

17. Ibid.

18. Betty J. Sternberg, commissioner, "Connecticut's 2004 Grads Continue Strong Performance on SAT," *News: Connecticut Department of Education*, Aug. 31, 2004.

19. Darling-Hammond, "Standards, Accountability, and School Reform," pp. 1047, 1049.

20. Gail L. Thompson, *Through Ebony Eyes: What Teachers Need to Know but Are Afraid to Ask about African American Students* (New York: Jossey-Bass, 2004), p. 35.

21. Ibid., pp. 65–68.

22. Palmer, *The Courage to Teach*, p. 10.

23. Edward Blanchard, personal communication, July 2, 2010.

24. Marilyn Frankenstein, "Critical Mathematics Education," in *Freire in the Classroom: A Sourcebook for Liberatory Teaching*, ed. I. Shor (Portsmouth, NH: Boynton/Cook, 1987), 186–87.

25. Robert Coles, *The Call of Stories* (Boston: Houghton Mifflin, 1989), pp. 11–12.

26. Tom Sito, "The Prism: A Profile of Dave Master," Animation World Network (AWN), May 20, 2008, www.awn.com/articles/profiles/prism-profile-dave-master.

27. Debra Feinstein, "When It Comes to Growing Talent, He's the Master," *Fast Company*, no. 5, Oct. 31, 1996.

28. Dave Master, personal communication, Jan. 9, 2010.

29. John Ramirez, personal communication, Jan. 24, 2010.

30. Mike Belzer, personal communication, Jan. 21, 2010.

31. Jay Parini, *The Art of Teaching* (New York: Oxford University Press, 2005), p. 55.

32. Marci Gray, personal communication, July 14, 2010.

33. Dave Master, personal communication, Jan. 8, 2011.

34. Dave Master, commencement speech, California Polytechnic Institute, Pomona, 2001.

35. Katherine Hanson and Bethany Carlson, *Effective Access: Teachers' Use of Digital Resources in STEM Teaching* (Newton, MA: Gender, Diversities, and Technology Institute at the Education Development Center, 2005).

36. N. Adelman et al., *The Integrated Studies of Educational Technology: Professional Development and Teachers' Use of Technology*, Tech. Rep. No. SRI Project P10474 (Menlo Park, CA: SRI International, 2002).

37. Daniel Pearl, Helene Cooper, and Mariane Pearl, *At Home in the World* (New York: Free Press, 2002), p. 178.

38. Ibid., p. 180.

39. B. J. Fogg, *Persuasive Technology: Using Computers to Change What We Think and Do* (San Francisco: Morgan Kaufmann, 2003), p. iii.

40. S. Chatterjee et al., "From Persuasion to Empowerment: A Layered Model, Metrics, and Measurement," paper presented at the Fourth Annual Persuasive Technology Conference, Copenhagen, Denmark, June 2010.

41. Ibid., p. 7.

42. Daniel Pink, *A Whole New Mind; Why Right-Brainers Will Rule the Future* (New York: Penguin, 2005).

43. Mihalyi Csikszentmihalyi, *Flow: The Psychology of Optimal Experience* (New York: Harper and Row, 1990), p. 71.

44. *Rising above the Gathering Storm: Energizing and Employing America for a Brighter Economic Future* (Washington, DC, National Academy of Sciences, 2007), p. 6.

CHAPTER 5: MENTORS AND HIGH EXPECTATIONS

Epigraph: E. Carpenter, F. Varley, and R. Flaherty, *Eskimo* (Toronto: University of Toronto Press, 1959).

1. National Center for Education Statistics, U.S. Department of Education, "Students Who Study Science, Technology, Engineering, and Mathematics (STEM) in Postsecondary Education," in *Stats in Brief*, Washington, DC, July 2009.

2. Ibid., p. 4.

3. Sheila Tobias, "Science Education Reform: What's Wrong with the Process?" *Change* 24 (May–June 1992).

4. Sheila Tobias, "What Makes Science Hard? A Karplus Lecture," *Journal of Science Education and Technology* 2, no. 1 (1993): 301.

5. Walter E. Massey, "A Success Story amid Decades of Disappointment," *Science* 258 (Nov. 1992): 1178–79.

6. Bill Long, "Confessions of a Non-Scientist," *Visions from Oregon Graduate Institute* 5, no. 3 (1989): 30–31.

7. F. H. T. Rhodes, *The Creation of the Future: The Role of the American University* (Ithaca, NY: Cornell University Press, 2001), p. 31.

8. Kenneth A. Feldman, "Research Productivity and Scholarly Accomplishment of College Teachers as Related to Their Instructional Effectiveness: A Review and Exploration," *Research in Higher Education* 26, no. 3 (1987): 227–98.

9. Paul Gray and David Drew, *What They Didn't Teach You in Graduate School: 199 Helpful Hints for Success in Your Academic Career* (Sterling, VA: Stylus, 2008).

10. Page Smith, *Killing the Spirit: Higher Education in America* (New York: Viking, 1990), p. 216.

11. Jeanne Nakamura and David J. Shernoff, with C. Hooker, *Good Mentoring: Fostering Excellent Practice in Higher Education* (San Francisco: John Wiley, 2009).

12. Ibid., pp. 209 and 96.

13. Richard Tapia, "Broadening Participation: Hiring and Developing Minority Faculty at Research Universities," *Communications of the ACM* 53, no. 3 (2010): 35.

14. Joseph Epstein, *Masters: Portraits of Great Teachers* (New York: Basic Books, 1981), pp. xi–xii.

15. Ibid, p. 61.

16. Smith, *Killing the Spirit*, p. 217.

17. Epstein, *Masters*, p. 176.

18. Smith, *Killing the Spirit*, p. 7.

19. Rhodes, *Creation of the Future*, p. 62.

20. Priscilla Gayle Harris Watkins, "Mentoring in the Scientific Disciplines: Presidential Awards for Excellence in Science, Mathematics Engineering Mentoring," UMI no. 3164230 (PhD diss., Claremont Graduate University, 2005), Abstract; hereafter cited by page number in the text.

21. P. Uri Treisman, "A Study of the Mathematics Performance of Black Students at the University of California, Berkeley" (PhD diss., University of California–Berkeley, 1985); hereafter cited by page number in the text.

22. Updated information about Dr. Treisman is taken from the Web site of the Dana Center at the University of Texas (www.utdanacenter.org/staff/uri-treisman.php).

CHAPTER 6: CLOSING THE ACHIEVEMENT GAP

Epigraph: Steven Strogatz, *The Calculus of Friendship: What a Teacher and a Student Learned about Life While Corresponding about Math* (Princeton: Princeton University Press, 2009), p. xii.

1. Martin V. Bonsangue and David E. Drew, "Long-Term Effectiveness of the Calculus Workshop Model" (report to the National Science Foundation, April 1992); hereafter cited by page number in the text.

2. Richard J. Light, *The Harvard Assessment Seminars: Explorations with Students and Faculty about Teaching, Learning, and Life* (Cambridge: Harvard University Graduate School of Education and Kennedy School of Government, 1992), p. 6.

3. David E. Drew, "The Impact of Reference Groups on the Several Dimensions of Competence in the Undergraduate Experience" (PhD diss., Harvard University, 1969).

4. Light, *Harvard Assessment Seminars*, p. 20; hereafter cited by page number in the text.

5. Fitzgerald Bramwell and Elinor Brown, "Louisiana Baccalaureate Origins of Doctorate Recipients in the Biological Sciences, Chemistry, Mathematics and Statistics, and Physics, 1979–2005," forthcoming.

6. Ibid, p. 6.

7. Diola Bagayoko, "Connecting Teaching and Mentoring to Learning: My Contributions to the American Academy," p. 48 in *Histoire de L'Afrique Noire: D'hier à Demain*, ed. J. Ki-Zerbo (Paris: Hatier, 2003). Hereafter cited by page number in the text.

8. Diola Bagayoko, "Philosophical Foundations for Systemic Mentoring at the Timbuktu Academy," *Science Career Magazine*, June 28, 2002, at http://sciencecareers.sciencemag.org/career_magazine/previous_issues; emphasis added.

9. Diola Bagayoko et al., "Basic and Research Training for the New Millennium: The Model of the Timbuktu Academy," *Journal of Materials Education* 24, nos. 1–3 (2002): 177–84.

10. Summary, "Numerical Results of the Timbuktu Academy, Southern University and A&M College in Baton Rouge (SUBR)," 2009, www.phys.subr.edu/TA/About_TA1.htm.

11. News Focus, *Science*, Oct. 27, 2006, p. 587.

12. Anthony Pullen, personal communication, Feb. 6, 2010.

13. Divine Kumah, personal communication, Feb. 6, 2010.

14. This section draws heavily upon evaluation reports that Martin Bonsangue and I prepared for the Houston LSAMP consortium.

15. Salina Vasquez-Mireles, "Correlating Mathematics and Science," *Mathematics Teaching in the Middle School* 15, no. 2 (2009): 100–107.

16. Armida Ornelas and Daniel G. Solorzano, "Transfer Conditions of Latina/o Community College Students: A Single Institution Case Study," *Community College Journal of Research and Practice* 28 (2004): 233–48.

17. Maria Teresa V. Taningco, Ann Bessie Mathew, and Harry P. Pachon, "STEM Professions: Opportunities and Challenges for Latinos in Science, Technology, Engineering, and Mathematics" (technical report, Tomas Rivera Policy Institute, 2008), p. 6.

CHAPTER 7: COLLEGE ACCESS AND THE STEM PIPELINE

Epigraph: Richard G. Lillard, *Desert Challenge: An Interpretation of Nevada* (1942), quoted in David Thomson, *In Nevada: The Land, the People, God, and Chance* (New York: Alfred A. Knopf, 1942), p. 275.

1. Andres Martinez, *24/7: Living It Up and Doubling Down in the New Las Vegas* (New York: Dell, 1999), p. 273.

2. Michael Ventura, "Las Vegas: The Odds on Anything," in *Literary Las Vegas: The Best Writing about America's Most Fabulous City*, ed. Mike Tronnes (New York: Henry Holt, 1995), pp. 179–80.

3. Ibid., pp. 175–76.

4. Advisory Committee to Examine Locating a Four-year State College in Henderson, "Committee's Report to the 2001 (71st) session of the Nevada Legislature," March 17, 2000, Henderson, Nevada. Elsewhere, the advisory committee report notes that a survey was conducted by Magellan Research of a sample of 400 residents of southern Nevada: "Over 70 percents of respondents agreed that 'Using Nevada tax dollars to build a state college in Henderson is a good idea.' Over 30 percent agreed with the statement that, 'If a state college existed in Henderson, someone in my household would likely attend classes there.' "

5. Ibid.

6. *A Technology Strategy for Nevada*, prepared by Battelle Memorial Institute, March 2001.

7. *Measuring Up, 2000: The State-By-State Report Card for Higher Education*, National Center for Public Policy and Higher Education, San Jose, CA, 2000.

8. P. Healy, "A Community College Meets Big Demand in Las Vegas," *Chronicle of Higher Education*, Feb. 23, 1996.

9. Michael O'Callaghan, "Where I Stand—Mike O'Callaghan: Punished for Successes," *Las Vegas Sun*, Nov. 2, 2000.

10. Howard R. Bowen, *Academic Recollections* (New York: Macmillan, 1988), p. 63.

11. F. H. T. Rhodes, *The Creation of the Future: The Role of the American University* (Ithaca, NY: Cornell University Press, 2001), pp. 89–90.

12. Harvard University, Faculty of Arts and Sciences, *Report of the Task Force on General Education* (Cambridge: Harvard University, 2007), p. 25. Online at http://isites.harvard.edu/fs/docs/icb.topic624259.files/report.pdf.

13. S. Balch and R. Zurcher, *The Dissolution of General Education: 1914–1993*, report prepared for the National Association of Scholars, Princeton, NJ, 1996.

14. David Denby, *Great Books: My Adventures with Homer, Rousseau, Woolf, and Other Indestructible Writers of the Western World* (New York: Simon and Schuster, 1997), p. 145.

15. J. Libby, "Design Contest for New College Launched," *Las Vegas Sun*, Jan. 5, 2001.

16. Nevada Board of Regents, *The Regents Review*, Special Issue, May 2001, p. 4.

17. J. Knight, "State College Startup Money Lost in Shuffle," *Las Vegas Sun*, June 5, 2001, p. 7B.

18. E. Vogel, "Henderson College: Regents Link Funding to Top Post," *Las Vegas Review-Journal*, June 16, 2001 p. 1B.

19. Michael O'Callaghan, "Where I Stand—Mike O'Callaghan: Real People Making News," *Las Vegas Sun*, June 21, 2001.

20. O'Callaghan, "Where I Stand—Mike O'Callaghan: Good Dreams of Better Education Shouldn't Be Interrupted," *Las Vegas Sun*, Mar. 1, 2002

21. President Maryanski passed away in July 2010.

CHAPTER 8: THE VALUE OF A COLLEGE EDUCATION IN THE GLOBAL ECONOMY

Epigraph: Quoted in Peter L. Bernstein, *Against the Gods: The Remarkable Story of Risk* (New York: John Wiley & Sons, 1996), p. 12.

1. David Drew, "Bank on What College Delivers, Not on What 'Apprentice' Presents," March 24, 2005, distributed by Knight-Rider and printed in about a dozen newspapers, including the *Providence Journal*, the *Detroit Free Press*, and the *Las Vegas Review Journal*.

2. Lee Halyard, personal communication.

3. Libby Parker, personal communication.

4. Nevadaworkforce.com, "Economy in Brief," Nov. 2009.

5. U.S. Department of Labor, Bureau of Labor Statistics, "Working in the 21st Century," www.bls.gov/opub/working/data/chart6.txt (year 2000); idem, "Employment Projections," www.bls.gov/emp/ep_chart_001.htm (year 2009).

6. Diane Stark Rentner and Nancy Kober, *Higher Learning—Higher Earnings: What You Need to Know about College and Careers*, Center on Education Policy (Washington, DC; American Youth Policy Forum, Sept. 2001), p. 5.

7. David Leonhardt, "The College Dropout Boom," *The New York Times*, May 24, 2005.

8. Gary Becker, *Human Capital: A Theoretical and Empirical Analysis, with Special Reference to Education* (Chicago: University of Chicago Press, 1993).

9. Scott L. Thomas, "Deferred Costs and Economic Returns to College Major, Quality, and Performance: Recent Trends," ASHE Annual Meeting Paper, Oct. 1998, p. 2.

10. Quoted in Howard R. Bowen, *Investment in Learning: The Individual and Social Value of American Higher Education* (Baltimore: Johns Hopkins University Press, 1977), p. 54.

11. David Littlejohn, *The Real Las Vegas: Life beyond the Strip* (New York: Oxford University Press, 1999), p. 213.

12. Quoted in Marc Reisner, *Cadillac Desert: The American West and Its Disappearing Water*, rev. ed. (New York: Penguin, 1993), pp. 120–21.

13. J. Ritter, "Vegas Drought May Wither Growth," *USA Today*, May 30, 2003, p. 3A.

14. Bureau of Labor Statistics, Local Area Unemployment Statistics, www.bls.gov/lau, Jan. 22, 2010.

15. Patrick Barta, "Wages Alter View of Las Vegas's Job Boom: Employment Data Prove Inadequate Gauge of Performance," *Wall Street Journal*, Oct. 16, 2000, p. A2.

16. Fox Butterfield, "As Gambling Grows, States Depend on It," *New York Times*, Mar. 31, 2005, pp. A1, A24.

17. Norman M. Klein, "Vegasthetics," in *The Magic Hour: The Convergence of Art and Las Vegas*, ed. Alex Farquharson (Ostfildern, Ger.: Hatje Cantz Publishers, 2001), p. 191. (In his notes to this paragraph, Klein cites Hubble Smith, "Nevada Economy: Slowdown Hits Las Vegas," *Las Vegas Review-Journal*, July 26, 2001; Jeff Simpson, "Casino Revenues: May Gaming Win Jumps," *Las Vegas Review-Journal*, July 11, 2001; and Cy Ryan, "Evidence of Las Vegas Casino Slowdown Mounts," *Las Vegas Sun*, Aug. 10, 2001.)

18. Dylan Rivera, "Timber Industry Begins to Show Signs of Stability," *The Oregonian*, Jan. 16, 2005.

19. Michael Milstein, "State Forests Can't Support Logging Plan," *The Oregonian*, May 15, 2005.

20. John Mitchell, "Sour Times in Sweet Home," *Audubon* 93 (Mar. 1991): 90.

21. Ibid., p. 91; emphasis added.

22. William H. Frey, "Charticle," *Milken Institute Review* 4, no. 3 (2002): 4.

23. Ibid, p. 7.

24. Robert B. Reich, *The Work of Nations: Preparing Ourselves for 21st Century Capitalism* (New York: Alfred A. Knopf, 1992), p. 172. Hereafter cited by page number in the text.

25. Robert B. Reich, "Harnessing Human Capital," *U.S. News and World Report*, Apr. 22, 1991.

26. Robert B. Reich, "The New Rich-Rich Gap," *Newsweek*, Nov. 28, 2005.

27. Peter Lyman, "The Library of the (Not-So-Distant) Future," *Change* 23 (Jan.–Feb. 1991): 40.

28. S. E. Berryman, quoted in U.S. Departments of Labor, Education, and Commerce, *Building a Quality Workforce* (Washington, DC: Office of Educational Research and Improvement, 1988), p. 11.

29. Brian Joyner, quoted in M. Schrage, "Statistics Skills Would Help U.S. Compete," *Los Angeles Times*, Mar. 14, 1991, p. D3.

30. H. Lee Barnes, *Dummy Up and Deal: Inside the Culture of Casino Dealing* (Lincoln: University of Nevada Press, 2002), p. x.

31. Richard Florida, *Cities and the Creative Class* (New York: Routledge, 2005), p. 24.

CHAPTER 9: SUPPORTING UNIVERSITY RESEARCH

Epigraph: Jeremy Bernstein, "The Faculty Meeting," *New Yorker,* Aug. 16, 1982, pp. 32–37.

1. National Science Board, *Science and Engineering Indicators, 2008,* 2 vols. (Arlington, VA: National Science Foundation, 2008), chap. 5.

2. Ibid.

3. Robert M. Berdahl, "Thoughts on the Current Status of American Research Universities" (presentation to the National Academy's Board on Higher Education and Work Force, Association of American Universities, 2009).

4. Adrian Wooldridge, "The Class of 2006: Why American Universities Will Lead the World," *Economist,* Nov. 18, 2006, p. 36.

5. David E. Drew, *Strengthening Academic Science* (New York: Praeger/Greenwood, 1985).

6. Richard A. Muller, "Innovation and Scientific Funding," *Science* 209 (Aug. 22, 1980): 880–83.

7. Henry Etzkowitz and Loet Leydesdorff, "The Dynamics of Innovation: From National Systems and 'Mode 2' to a Triple Helix of University–Industry–Government Relations," *Research Policy* 29 (2009): 116.

8. Randy L. Haupt, "The Industry-University Intellectual Property Paradox," *IEEE Antennas and Propagation Magazine* 46, no. 2 (2006): 134.

9. Ibid.

10. Nikolaos A. Patsopoulos, Apostolos A. Analatos, and John P. A. Loannidis, "Origin and Funding of the Most Frequently Cited Papers in Medicine: Database Analysis," *British Medical Journal* 332 (Mar. 2006): 1061–64.

11. S. Nadis, "U.S. Concern Grows over Secrecy Clauses," *Nature* 398 (1999): 359.

12. David E. Drew, *Science Development: An Evaluation Study* (Washington, DC: National Academy of Sciences Press, 1975); National Board on Graduate Education, *Science Development, University Development, and the Federal Government,* National Board on Graduate Education Technical Report no. 4 (Washington, DC: National Academy of Sciences Press, 1975).

13. Rudy Abramson, "Patron on the Potomac: The National Science Foundation," *Change* 3 (1971): 38–43.

14. This descriptive information about the initial design of the Experimental Program is based upon NSF program documents.

15. David E. Drew, *Strengthening Academic Science* (New York: Praeger, 1985).

16. Quotations and current EPSCoR data were acquired from NSF, Experimental Program to Stimulate Competitive Research (EPSCoR) Web site, www.nsf.gov/epscor.

17. All three of these projects were reported on the NSF Web site, at www.nsf.gov/discoveries: "Cheaper Plastic Solar Cells in the Works," Jan. 13, 2009; "New Coating

Could Prevent Infection from Surgical Tools and Implants," Sept. 7, 2007; and "Life Can Be a Strain," Feb. 21, 2007.

18. EPSCoR/IDeA Foundation Web site, www.epscorfoundation.org.

19. *Rising above the Gathering Storm: Energizing and Employing America for a Brighter Economic Future* (Washington, DC, National Academy of Sciences, 2007).

CONCLUSION

1. John Wooden, with Steve Jamison, *Wooden: A Lifetime of Observations and Reflections On and Off the Court* (New York: McGraw-Hill, 1997).

2. James Fallows, *More Like Us* (Boston: Houghton Mifflin, 1990).

3. Nassim Nicholas Taleb, *The Black Swan: The Impact of the Highly Improbable* (New York: Random House, 2007).

4. Andrew Greeley, "They Leap from Your Brain, then Take over Your Heart," in *Writers on Writing: More Collected Essays from "The New York Times,"* vol. 2 (New York: Henry Holt, 2003), p. 86.

5. Ray Kurzweil, *The Age of Spiritual Machines: When Computers Exceed Human Intelligence* (New York: Penguin Group USA, 2000).

6. E. F. Schumacher, *Small Is Beautiful: Economics As If People Mattered* (New York: Harper & Row, 1973). Hereafter cited by page number in the text.

7. E. Cornish, "Think Small," *The Futurist*, Dec. 1974, p. 276.

8. S. Love, "We Must Make Things Smaller and Simpler: An Interview with E. F. Schumacher," *The Futurist*, Dec. 1974, p. 282.

9. David A. Peters, "Boomers, Bloomers, and Zoomers: The Aerospace Partnership Lives On," *Vital Speeches of the Day* 62, no. 12 (1999): 379.

10. "Opening Day Rosters Feature 229 Players Born Outside the U.S.," press release, Apr. 6, 2009, Major League Baseball Web site, at http://mlb.mlb.com/news/press_releases/press_release.jsp?ymd=20090406&content_id=4139614&vkey=pr_mlb&fext=.jsp&c_id=mlb.

11. Sheila Tobias, "Science Education Reform: What's Wrong with the Process?" *Change* 24 (May–June 1992): 16–17; emphasis added.

12. Richard Stern, "Wissler Remembers," in *Packages: Stories by Richard Stern* (New York: Coward, McCann & Geoghegan, 1980), pp. 11–12.

13. Kent Kavanaugh, testimony before the Subcommittee on Postsecondary Education of the U.S. House of Representatives, *Math, Science, and Engineering Education: A National Need: Hearing before the Subcommittee on Postsecondary Education of the Committee on Education and Labor*, 101st Cong., 1st sess., Kansas City, MO, May 1, 1989.

14. James W. Stigler and James Hiebert, *The Teaching Gap* (New York: Free Press, 1999).

15. Jean Lipman-Blumen, *The Connective Edge: Leading in an Inter-dependent World* (San Francisco: Jossey-Bass, 1996), p. 253.

16. Ibid., p. 4.

17. Elie Kedourie, "Politics and the Academy," *Commentary* 94, no. 2 (1992): 55.

18. Laura Andersson, "But Why?" *Visions* 5, no. 3 (1989): 31.

APPENDIX

1. T. Hussen, ed., *International Study of Achievement in Mathematics* (New York: John Wiley, 1967).

2. L. C. Comber and J. P. Keeves, eds., *Science Education in Nineteen Countries: An Empirical Study* (New York: John Wiley, 1973); D. Walker, *The IEA Six-Subject Survey: An Empirical Study of Education in Twenty-One Countries* (New York: John Wiley, 1976); R. M. Wolf, *Achievement in America: National Report of the United States for the International Educational Achievement Project* (New York: Teachers College Press, 1977).

3. Wolf, *Achievement in America.*

4. T. Husen, in Comber and Keeves, *Science Education in Nineteen Countries,* p. 10.

5. Ibid., p. 17.

6. Ibid.

7. Wolf, *Achievement in America,* p. 188; emphasis added.

8. Curtis C. McKnight et al., eds., *The Underachieving Curriculum: Assessing U.S. School Mathematics from an International Perspective; A National Report on the Second International Mathematics Study* (Champaign, IL: Stipes, 1987), p. vii.

9. Ibid.

10. Ibid., p. xii.

11. Ibid., pp. 85, 95.

12. International Association for the Evaluation of Educational Achievement (IEA), *Science Achievement in Seventeen Countries: A Preliminary Report* (New York: Pergamon, 1988), p. 20.

13. Ibid., p. 4; emphasis added.

14. Ibid., p. 68.

15. J. Eichinger, "Science Education in the United States: Are Things as Bad as the Recent IEA Report Suggests?" *School Science and Mathematics* 90 (Jan. 1990): 38.

16. IEA, *Science Achievement in Seventeen Countries,* p. 20.

17. A. E. LaPointe, N. A. Mead, and J. M. Askew, *Learning Mathematics,* report no. 22-CAEP-01 (Princeton, NJ: Educational Testing Service, 1992), p. 4.

Index

Page numbers in *italics* indicate figures and tables.

About the Author

David E. Drew holds the Joseph B. Platt Chair in the Management of Technology at the Claremont (CA) Graduate University. His principal appointment is in Education. He also has cross-appointments in Management, Psychology, and Mathematics. For 10 years Mr. Drew served as dean of the School of Educational Studies.

Prior to joining the CGU faculty, he held senior research positions at the Rand Corporation, the National Research Council, and the American Council on Education. Earlier he held a research faculty position at Harvard University, from which he received his PhD, and served as head applications programmer at the Harvard Computing Center.

He is the author of more than 150 publications, including 9 books, about the improvement of mathematics and science instruction at all levels of education, the development and evaluation of effective undergraduate programs, building strong university research programs, and health education.

Recently the National University of Singapore celebrated its 100-year anniversary and hosted an international conference about education and globalization. Mr. Drew was invited to give the keynote speech.

In the past few years, he has been a consultant or advisor to the senior leadership of the Nevada higher education system, the College Board, the Los Angeles Police Department, the Tibetan Buddhists exiled in India (pro bono), and Patagonia, an international clothing company.